49

Stem Cells, Human Embryos and Ethics

Lars Østnor
Editor

Stem Cells, Human Embryos and Ethics

Interdisciplinary Perspectives

 Springer

Editor
Lars Østnor
MF Norwegian School of Theology
Oslo, Norway

ISBN: 978-1-4020-6988-8   e-ISBN: 978-1-4020-6989-5

Library of Congress Control Number: 2008922457

Cover picture: The cover image is a graphic art picture called "Hyss" by the Norwegian artist John Thørrisen. Used by permission of the artist.

Printed on acid-free paper

9 8 7 6 5 4 3 2 1

springer.com

Foreword

The successful isolation of human embryonic stem cells by Dr. Jamie Thomson in 1998 has lead to subsequent derivation of over a hundred additional lines. There has been a significant improvement in the efficiency of derivation and a number of variations on the basic methodology have been described. These include isolation of parthenogenetic lines, isolation from the morula stage embryo and isolation from later stages of embryonic development. Equally importantly there have been reports of successful reprogramming of adult cells into embryonic stem cells and derivation of epiblast like cells. Thus there are possibilities of obtaining cells that may not make teratomas or contribute to the germline or be classified as living embryos and may even bypass the ethical issues raised by oocyte donation while still retaining many of the characteristics of a pluripotent stem cell.

These successes and the ethical and social issues that manipulating ones own genome raise have lead to a fierce debate that has not simply been confined to scientists or ethicists but has spilled into the mainstream and in some cases been politicized. Each group has taken its own extreme position and in the case of the United States individual states have taken positions that differ from the official federal policy. National-level review is required in only a few countries (e.g., the Human Fertilization and Embryology Authority in the UK) and in the US, the idea of national review is still under consideration.

Despite the ongoing ethical debate research under current guidelines has proceeded relatively rapidly and attempts to translate basic stem cell research into treatments for neurological diseases and injury are well underway. Although there are perhaps more policy and ethical discussion papers and reviews on ESC then there are research reports one can perhaps make a case that stem cell science is proceeding faster than the social debate concerning the ethical integrity of the research and the protection of potential human subjects in the research.

Indeed, in the United States (US), the Food and Drug Administration (FDA) has approved an Investigational New Drug application (IND) for using human central nervous system stem cells, isolated from fetal brain tissue, in clinical trials testing a treatment for Batten disease, a fatal inherited disorder of the nervous system and unconfirmed reports indicate that the FDA is reviewing an application from Geron to use hESC derived oligodendrocytes to treat spinal cord injury and press reports from India, China and other countries of ESC derive cell transplants.

The European Union perhaps represents a region where opinions differ markedly from country to country. There are complete bans in some countries, relatively liberal regulations in others and some intermediate (although different) compromises in most other countries. It is in this context that a book such as this one is welcomed. It is a comprehensive collection of chapters covering all major aspects of the ethical debate presented in an unbiased way. It is a fitting culmination of the numerous research projects on bioethics that Dr. Lars Østnor has either lead or participated in. Dr. Lars Østnor is to be commended for the effort he has taken in recruiting a stellar group of contributors that represent the continuum of opinion on ESC in Europe. It deals with the topic of the moral status of human embryos with special regard to stem cell research and therapy. The book contains contributions from top professionals within biology, medicine, philosophy and theology. The different chapters are the result of a major international research project in the years 2005 and 2006 with participants from USA, United Kingdom, The Netherlands, Germany and the Nordic countries. Among the contributors to the project and among the authors of the book are Professor William B. Hurlbut from Stanford University, Professor LeRoy Walters from Kennedy Institute of Ethics, Georgetown University and Professor Dagfinn Føllesdal from University of Oslo and Stanford University.

There are not many truly comprehensive books on these contentious issues where one can examine the opinions of both religious and scientific scholars side by side. To see this in one book where the topics are covered in a forthright and non contentious fashion is even rarer. Dr. Østnor has done an exceptional job and I hope the readers will find this book as useful as I do.

Mahendra Rao

Introduction

This book is a multidisciplinary study investigating the field of embryonic stem cells from different professional perspectives. The book has both a biological/medical perspective, dealing with new technological possibilities in medical research and putative clinical therapy, as well as an ethical perspective, including philosophical and theological approaches focusing on the question of the moral status of human embryos.

The researchers involved in the study represent different scientific, philosophical and theological positions, and different views concerning the ethical problems at issue. The idea is that cooperation in such a group creates a critical and self-critical dialogue where differing opinions and evaluations can be reexamined.

There are several reasons for writing this book: (a) The stem cell field is a very 'hot' and promising field of research in many countries around the world. (b) At the same time it is a controversial field, especially with regard to some of the ethical implications. Ethical aspects of stem cell research are highly debated among people from biology, medicine, law, philosophy, theology etc. (c) There are different regulations concerning stem cell research within the legislation of various countries. (d) The public discussions among citizens and among politicians are to a large extent dividing the populations of many countries in the West and the East. (e) On this background I decided to bring together professionals from different disciplines involved in the topic in order to start an international and interdisciplinary work concerning some of the burning ethical questions raised by stem cell research and eventual therapy. This work was carried out as a two year research project financially supported by the Norwegian Research Council, with the title "The moral status of human embryos with special regard to stem cell research and therapy". The reason for this approach was that the international debate has been especially concentrated on the use of embryonic stem cells. (f) Selected presentations from this project have later been rewritten with a view to publishing these final texts in a common book. On the one hand the contributions give comprehensive and updated information on the current situation within stem cell research, and on the other hand they give a presentation and an evaluation of the ethical argumentation related to the field. (g) Finally, I am of the strong opinion that the debate going on in many countries concerning stem cell research, and especially the use of human, embryonic stem cells, will profit substantially from a sufficient overview of the different aspects relevant for an ethical evaluation.

The book differs from others in the field in two ways: (a) It gives interdisciplinary perspectives from several relevant professional fields, and (b) it has a specific focus by concentrating on one main problem. This problem is further elaborated by the raising of various biological, medical and ethical sub-problems.

The main problem at the center of this book is: What is the moral status of human embryos with regard to the use of embryonic stem cells in scientific research and clinical therapy, and what are the weaknesses and strengths of various opinions and positions when they are critically evaluated?

This involves some biological/medical sub-problems: What is the state of biological and medical research regarding embryonic stem cells? What are the prospects for the future regarding the therapeutic use of embryonic stem cells? And what about the possibilities of other sources of stem cells?

Further, the main problem also involves several ethical sub-problems: What is moral status and what characterizes the moral status of human embryos? What is the moral status of human embryos according to different philosophical and theological traditions and what are the weaknesses and strengths of the various traditions? Are there ethically relevant differences between the various ways in which embryonic stem cells are made available, including therapeutic cloning? What are the main ethical problems and dilemmas generated by research with and therapeutic use of embryonic stem cells? How do we balance respect for embryos over against the need to advance life-saving and suffering-reducing medical progress?

Chapter 1 of the book contains a common statement from the Norwegian project group. It is a summary of the discussions during the project period and in the group. It gives priority to some aspects of the problem that are interesting from an ethical point of view. Chapters 2–6 bring several contributions from top professionals within science. Chapters 7–9 contain contributions from philosophy and theology with regard to the different social and political aspects of the stem cell field, the public discussions, and the state legislations regarding stem cell research. Chapters 10–14 give philosophical contributions concerning some burning ethical problems arising from stem cell research. In chapters 15–17 three theologians present and evaluate the theological argumentation in the human stem cell debate and in part also give their own reflections regarding certain central ethical aspects in the stem cell field.

The various chapters all together give a comprehensive, multifaceted and balanced treatment of the subject of the book. A few of the chapters touch in part upon some common aspects of the problem discussed, but they do this from different professional angles and because of that they complement each other rather than repeat the same perspectives.

This book is a multi-author or edited book. It is not a conference proceeding. The authors have been selected because of their professional competence, many of them being respected scholars at a top international level. They have also been chosen in order to give an updated contribution from their own disciplines and enlighten different, defined aspects of the common theme.

This book is written for several audiences: (a) a range of scholars or professionals working with stem cell research and the ethical questions arising from this field:

people from biology, medicine, law, philosophy, theology etc.; (b) advanced and graduate students within the same professional disciplines; and (c) politicians and the general public interested in the burning ethical problems which at present are debated and need social and legal regulations.

Being the editor of the book I would like to express my gratitude to the authors for using their insights and time in giving substantial contributions! Thanks to the Norwegian Research Council for financial support, to their Senior Advisor Helge Rynning for practical assistance and to MF Norwegian School of Theology and President Vidar L. Haanes for including this work among the research projects of this institution! Thanks also to Professor Mahendra Rao for his willingness to write a preface for the book and to Torhild Øien for her careful preparation of the manuscripts for publication! Finally, but not least, a thank you to Associate Editor Max Haring at Springer Publishers for his positive engagement and excellent cooperation during the publication process!

Oslo   Lars Østnor
October 2007

Contents

Contributors

The editor **Lars Østnor** is Professor of systematic theology with special regard to ethics at MF Norwegian School of Theology, Oslo. He was the founder of and coordinator for The Nordic Theological Network for Bioethics 1992. He is a member of Societas Ethica (a European organization for ethicists) 1998 – and a member of the board of The Nordic Society of Theological Ethics 2001. He has published five books and edited many books, among them: *Bioetikk og samfunn (Bioethics and Society)*. Oslo 1995; *Bioetikk og teologi (Bioethics and Theology)*. Oslo 1996; *Bioetikk, evtanasi og omsorg (Bioethics, Euthanasia and Care)*. Oslo 1997; *Bioethics and Cloning*. Oslo 1998; and *Etisk pluralisme i Norden (Ethical Pluralism in the Nordic Countries)*. Kristiansand 2001.

Mahendra Rao is the Vice President of Research in Stem Cells and Regenerative Medicine for Invitrogen Corporation, Carlsbad, California. He holds an M.D. from Bombay University and a Ph.D. in developmental neurobiology from California Institute of Technology. He has been teaching at Johns Hopkins University School of Medicine, the National Centre for Biological Sciences in Bangalore and the University of Utah School of Medicine. He has published more than 100 papers on stem cell research and several books, among them *Stem Cells and CNS Development* and *Neural Development and Stem Cells*.

Steinar Funderud has been heading the stem cell research at the Comprehensive Cancer Center, The Norwegian Radium Hospital and is at present chairman of the board for tumor stem cell center. He has been Professor in tumor immunology at the Faculty of Medicine, University of Oslo. His doctoral thesis is *Mechanisms of DNA Replication in Physarum Polychepalum*. University of Tromsø, Norway, 1978. His speciality is hematopoietic stem cells.

Ole Johan Borge is Senior Advisor at The Norwegian Biotechnology Advisory Board. He has a Ph.D. in stem cell research from Lund University. His doctoral thesis is *Maintenance and Commitment of Hematopoietic Progenitors: Role of flt3, c-kit and c-mpl*. Lund University, Sweden, 1999. Borge is the co-editor of many publications within the field of biotechnology.

Iver Langmoen M.D., Ph.D., is Professor at the Department of Neurosurgery, Ullevål University Hospital in Oslo, and teaches at the Faculty of Medicine,

University of Oslo. He is also head of the Vilhelm Magnus Laboratory for Neurosurgical Research, which is an integrated research unit for the two neurosurgical departments at the University of Oslo. From 1997 to 2005 he was Professor of Neurosurgery at Karolinska Institutet, Stockholm, Sweden. His doctoral thesis is *Synaptic Mechanisms in Hippocampal Pyramidal Cells*. University of Oslo, Norway, 1981.

Håvard Ølstørn, M.D., and **Mercy Varghese**, M.D., work at the Department of Neurosurgery at Ullevål University Hospital, and are Ph.D. students in Professor Langmoen's research group. **Morten C. Moe**, M.D., Ph.D., is a postdoctoral fellow in Langmoen's group, and is working at the Department of Ophthalmology at Ullevål University Hospital.

Joel C. Glover Ph.D. is Professor at the Department of Physiology, Institute of Basic Medical Sciences, University of Oslo and is also affiliated with the Norwegian Center for Stem Cell Research and the Cancer Stem Cell Innovation Center. He received his doctorate from the University of California, Berkeley in 1984. His research focuses on spinal cord development and regeneration, the potential of human somatic stem cells for neural differentiation, and the evolution of the vertebrate brain.

William B. Hurlbut M.D. is Consulting Professor at The Neuroscience Institute at Stanford, Stanford University Medical Center, USA. He is a member of The President's Council on Bioethics, Washington, DC. He is especially known worldwide for his scientific publications concerning Altered Nuclear Transfer (ANT) – a specific technique for deriving human embryonic stem cells without creating and destroying an embryo.

Professor **LeRoy Walters** is a theologian and philosopher at the Kennedy Institute of Ethics and in the Department of Philosophy at Georgetown University, Washington, DC. He is coauthor of *The Ethics of Human Gene Therapy,* Oxford University Press, New York 1997 and coeditor of *Contemporary Issues in Bioethics,* 7th ed., Wadworth, Belmont, CA, 2008, *Source Book in Bioethics,* Georgetown University Press, Washington, DC, 1998, and the annual *Bibliography of Bioethics,* Kennedy Institute of Ethics, Washington, DC, 1975–2007.

Egbert Schroten received his Ph.D. from Utrecht University in 1970 (thesis on the meaning of corporeal existence). He was lecturer in philosophy of religion from 1969 to 1987 at the Faculty of Theology of Utrecht University. From 1987 to 2004 he was Professor for Christian ethics at the same faculty and director of the University Centre for Bioethics and Health Law. From 1994 to 2001 he was a member of the European Group on Ethics in Science and New Technologies to the European Commission. He has also been the Moderator of the Working Group on Bioethics of the Church and Society Commission of the Conference of European Churches (CEC).

Theo A. Boer is Associate Professor of ethics at the Protestant Theological University in Utrecht and Associate at the Ethics Institute of Utrecht University.

He studied theology and ethics in Utrecht and Uppsala. He publishes both about fundamental theological ethics and about various applied ethical issues.

Dagfinn Føllesdal studied science and mathematics in Oslo and Göttingen before going to Harvard to study with Quine. After his Ph.D. in 1961 he taught at Harvard and then in Oslo (1967–99) and also at Stanford since 1968, where since 1976 he has been C.I. Lewis Professor of Philosophy. Publications in philosophy of language and on phenomenology. Editor, *The Journal of Symbolic Logic*, 1970–82. Member of American, English, German and Scandinavian Academies of Science. Former President of the Norwegian Academy of Science. Headed the Norwegian Research Council's ethics program 1992–2002.

Øyvind Baune is Professor in philosophy at the Department of Philosophy, Classics, History of Art and Ideas, University of Oslo. He has worked within philosophy of mathematics, natural science and language, and in later years mainly in ethics. He has been member of The National Committee for Medical Research Ethics in Norway and been part of an EU-project in the development of assessments tools for agriculture and food production. He has published articles within these areas and written books within the fields of scientific methodology, logic and theory of argumentation.

Sir Anthony Kenny is Professor emeritus from the Faculty of Philosophy, University of Oxford, Great Britain. He has been Master of Balliol College, Oxford, President of the British Academy and Chairman of the Board of the British Library. Since the 1960s he has written a great number of influential books on the history of philosophy. Among them is *A Brief History of Western Philosophy*. Blackwell Publishers. Malden, MA, 1998. Other books are studies in Aristotle, Aquinas, Descartes, Wittgenstein etc.

Ludger Honnefelder is Professor emeritus of philosophy at the University of Bonn, Germany and Director of the Institute for Science and Ethics at the University in Bonn, Germany. Since 2005 he is Professor of philosophy of religion and Roman Catholic world view (Weltanschauung) at the Department of Theology, Humboldt University, Berlin, Germany. His doctoral thesis at Friedrich-Wilhelms-Universität, Bonn is *Ens inquantum ens: der Begriff des Seienden als solchen als Gegenstand der Metaphysik nach der Lehre des Johannes Duns Scotus*. Aschendorff. Münster 1979. He has published several books on metaphysics, ethics (including applied ethics) and the history of medieval philosophy and its reception in early modern philosophy.

Otfried Höffe is Professor of philosophy at the University of Tübingen, Germany (Philosophisches Seminar), Founder and Director of the Research Center of Political Philosophy and permanent guest Professor at the University of St. Gallen. His professional interests cover Aristotle, Kant, moral and political philosophy, applied ethics and epistemology. His nearly two dozen books have been translated into 20 languages. Among his latest books are *Democracy in an Age of Globalisation*. Springer NL, Dordrecht 2007, *Kants Kritik der reinen Vernunft. Die*

Grundlegung der modernen Philosophie. C.H. Beck Verlag. München 4th ed. 2004 (engl. *Kant's Critique of Pure Reason*, Springer NL, Dordrect, forthcoming 2008) and *Wirtschaftsbürger, Staatsbürger, Weltbürger. Politische Ethik im Zeitalter der Globalisierung*. C.H. Beck Verlag. München 2004.

Gunnar Heiene is Professor of theological ethics at MF Norwegian School of Theology, Oslo. His doctoral thesis is from this institution and has the title *Den menneskelige stat. Antropologi og politikk hos Eivind Berggrav (The Human State: Anthropology and Politics in Eivind Berggrav* (Norwegian bishop in Oslo during the second world war). Oslo 1991. Later he has published books and articles from different areas of theological ethics, especially within fundamental ethics, family and political ethics. Together with colleagues he has also published several text books for theological studies.

Monika Bobbert, Dr. theol. and dipl. psych., is working at the Department of Medical Ethics at the Institute for the History of Medicine, the Medical Faculty of the University of Heidelberg, Germany. Her doctoral thesis is *Patient Autonomy and Nursing. Foundation and Application of a Moral Right*. Her speciality is bioethics and other fields of applied ethics.

Chapter 1
The Moral Status of Human Embryos with Special Regard to Stem Cell Research and Therapy

Øyvind Baune, Ole Johan Borge, Steinar Funderud, Dagfinn Føllesdal, Gunnar Heiene, and Lars Østnor*

Abstract This chapter contains a common statement from the Norwegian project group. The statement gives an introduction into embryo development and the stem cell field, including the question of different sources for human embryonic stem cells. It delivers a survey of some of the argumentative concepts within philosophical and theological debates regarding the moral status of human embryos and evaluates the relevance, strengths and weaknesses of the arguments. The statement also deals with ethical-normative elements connected to human biological life and ethical norms with relevance for research and clinical therapy. Finally, it includes a reflection about the ethical dilemma between medical progress on one hand and respect for embryos on the other. The conclusion outlines the profile of the two different positions represented in the group.

Keywords The stem cell field, ethical traditions, normativity, dilemmas, alternative sources

1.1 Introduction

The following statement sums up results of a two year research project with the title 'The moral status of human embryos with special regard to stem cell research and therapy'. The project was financially sponsored by the Norwegian Research Council. This interdisciplinary project had a steering group consisting of Professor Øyvind Baune, University of Oslo (philosophy), Senior Advisor Ole Johan Borge, Ph.D., The Norwegian Biotechnology Advisory Board (biology), Professor Steinar Funderud, Rikshospitalet-Radiumhospitalet HF (biology), Professor Dagfinn Føllesdal, University of Oslo and Stanford University, USA (philosophy), Professor Gunnar Heiene, MF Norwegian School of Theology (theology) and Professor Lars Østnor,

* MF Norwegian School of Theology, Box 5144 Majorstua, 0302 Oslo, Norway.
e-mail: Lars.Ostnor@mf.no

MF Norwegian School of Theology (theology, head of the group). Among active participants in the project were also Professor Daniel Callahan, The Hastings Center, New York, Professor LeRoy Walters, Kennedy Institute of Ethics, Washington, Professor Egbert Schroten and Dr. Theo A. Boer, both from Ethics Institute, Utrecht University. The participants represent different milieus and various positions.

During the years 2005 and 2006 four workshops and conferences were arranged within the framework of the project. Each time 20–30 persons from many professional fields were invited to give lectures, responses or contributions to the debates. An open, final conference had attendees from research, politics, media, the general public etc. A doctoral student working on her Ph.D. thesis is also included in the project. In addition there have been several meetings in the project group. This statement is a summary of professional discussions during the workshops and in the group. Its aim is to survey those aspects of the problem that are of particular ethical relevance. It does not consider legal questions.

1.2   The Topic

The expression 'moral status' has almost become a technical term within several disciplines for the following: that something (human, animal, plant, etc.) has some form of moral status implies that we as moral agents have ethical obligations towards it. Those who have moral status must, from the viewpoint of agents, be protected by certain ethical norms.

When we raise the problem of the moral status of human embryos, we are concerned on the one hand with what *rights* they have to the protection of life, body, health etc., and on the other hand what *obligations* moral agents have towards them in the form of preserving these goods.

Within biology and medicine, one distinguishes at times between the various phases in early human development: fertilized ovum (zygote), morula, blastocyst and fetus. The number of phases and decisions concerning terminology vary dependent on empirical, ethical and legal factors. In this statement we use '*embryo*' inclusively to signify human life from fertilization to the eighth week of life. The addition of '*human*' signifies that we are speaking of the human species.

Stem cells are undifferentiated, self-renewing cells with the potential to produce specialized, differentiated cells. There are various kinds of stem cells, based on where they are found in the human organism. *Human, embryonic stem cells* (hES cells) can be derived from the blastocyst ca. 5–10 days after fertilization. Because they have the ability to produce every cell type in the human body, they are often called pluripotent. The possibility of deriving hES cells from an embryo raises the important and difficult question of defining ethically acceptable actions toward human life at this phase of development.

The interest of *science* in being able to make use of hES cells is especially related to three objectives: (a) better knowledge of human biological development at an early stage of its life (basic research), (b) testing of both new and existing compounds/drugs, and (c) the cultivation of various types of cells, tissues, and perhaps even organs.

Research on hES cells may open up for new *clinical treatments* of illnesses and injuries. Active research is today being carried out in many countries, but no one has yet attained any recognized therapy of the serious illnesses one hopes to be able to fight.

1.3   Embryo Development and the Stem Cell Field

1.3.1   Embryo and Fetal Development

Sperm penetrating the eggshell (*zona pellucida*) initiates fertilization and the creation of the *zygote* – a cell containing genetic material from both parents. The *zygote* divides and generates two *blastomeres*. As the blastomeres continue to divide approximately every 20 hours, they increase in number but become smaller for each division. After three days a ball of about 16 cells, called a *morula*, appears. About four days after fertilization a cavity forms within the morula and the structure is then called a *blastocyst*. At this point, the cells have started to differentiate and are no longer considered totipotent. A totipotent cell has the intrinsic ability to generate a fetus if implanted in a uterus. As the cells continue to divide, the pressure on the eggshell increases, and after 5–6 days the blastocyst 'hatches' (leaves the eggshell) and begins to find its place in the uterus. The implantation of the embryo in the uterus is completed at day 12. At day 13 the first signs of a placenta and an umbilical cord can be seen. The first visual sign of organ formation, named the primitive streak, appears around day 14. Splitting of the embryo, generating monozygotic twins, can happen until the presence of the primitive streak.

In the following days and weeks, organ formation continues and all the main organs and structures of the developing embryo are present after approximately 50 days. This is also the time at which the first signs of brain activity can be measured and at which the embryo makes spontaneous movements. By definition the embryo period ends eight weeks/two months (56–60 days) after fertilization, at which time the embryo is 23–26 mm in size.

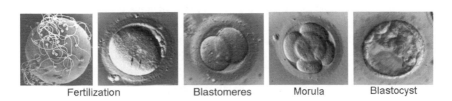

Fertilization   Blastomeres   Morula   Blastocyst

Fig. 1.1 Photo: David Epel, Stanford, CA, USA and the Norwegian Biotechnology Advisory Board

1.3.2 Stem Cells Basics

Results published in the late 1950s and early 1960s established the concept of stem cell when it was demonstrated that lethally irradiated animals could be rescued by transplantation with bone marrow cells from a donor animal.

There are generally two main types of stem cells characterized according to their potential to differentiate: pluripotent and multipotent.

1.3.2.1 Pluripotent Stem Cells

By definition, pluripotent stem cells have the potential to differentiate into all cell types in the adult body. However, pluripotent stem cells cannot form an entire individual if implanted in a uterus, because they are unable to give rise to extraembryonal tissues (like the placenta) essential for fetal development.

There are currently demonstrated three sources of pluripotent stem cells: early embryos, fetuses and teratocarcinomas (a rare form of cancer). Stem cells derived from the human embryo are called embryonic stem cells (hES cells). Typically, hES cells are derived from the inner cell mass of blastocysts (5–8 days after fertilization), but hES cells have also been isolated at the 8/16-cell stage (2–4 days after fertilization). Pluripotent stem cells isolated from fetuses are termed EG cells (embryonic germ cells) and pluripotent stem cells from teratocarcinomas are termed EC cells. Whether pluripotent stem cells also exist naturally in adult individuals is highly debated and no definite proof has been given to date. Recent data may however indicate that both the bone marrow of adults and the amniotic fluid potentially contain pluripotent like stem cells.

1.3.2.2 Alternative Sources for hES Cells

hES cells have, as described above, generally been looked upon as *one* cell type with *one* set of characteristics and whose only source is the developing, healthy embryo. However, recently a number of alternative sources have been presented. Some of these are merely of theoretical interest, whereas others might represent attractive alternatives to healthy embryos.

Briefly, these alternatives include:

- Embryos incapable of further development. After in vitro fertilization a high fraction of the embryos stop at various phases of development. In relation to fertility treatment, these embryos are simply discarded. It has recently been demonstrated that there are viable cells within these embryos and that they may be used as sources for developing viable hES cells.
- Single blastomeres withdrawn from the embryo without destruction. Based on experience with Preimplantation Genetic Diagnosis (PGD) we know that single cells can be withdrawn from developing embryos. The withdrawn cells may further be used to establish hES cells.

- Somatic cell nuclear transfer (SCNT). In this technique a somatic cell nucleus is inserted into an enucleated egg cell. If such an egg cell is stimulated, the cell starts to divide and progress into a blastocyst from which hES cells may be developed.
- Altered nuclear transfer (ANT). This technique is a variant of SCNT where the transferred nucleus is altered so that no blastocyst develops. The biological entity formed lacks the capacity of an embryo, but hES cells can in principle be derived.
- Redifferentiation of somatic cells to a state enabling the derivation of hES cells. It can be assumed that there are a limited number of genes responsible for defining any given cell type. Results indicate that it might be possible to manipulate adult, somatic cells to convert to a state resembling the hES cell state.

1.3.2.3 Multipotent Stem Cells

Multipotent, somatic stem cells are during adult life responsible for maintaining homeostasis in every tissue, organ and cell system. Multipotent stem cells are currently considered tissue specific. They can only produce tissue of the same type. The stem cells in bone marrow and brain are to date the ones best characterized. Multipotent stem cells can also be isolated from the umbilical cord immediately after birth.

1.3.3 Usage of Stem Cells

All types of stem cells are highly attractive study objects due to their undifferentiated state, prolifer--- ---ability, and ability to differentiate into all cell types
--- lition to their key role in developmental biology and
so increasingly acknowledged as a main factor in the
rime target in cancer therapy.
few cases multipotent stem cells tend to lose their
are withdrawn from their natural environment.
tipotent stem cells in a defined culture system. hES
pagated and stimulated to differentiate in culture.
potential to differentiate into all cell types and are
issue system.
also be useful as a tool for drug discovery and

d blood transplantation, utilizing stem cells, is
of patients with bone marrow disorders and
going clinical trials using adult stem cells are
Although stem cells are currently being used
re is considerable hope that stem cells in the
of a wide array of human disorders. In par-
-- a complete lack, or a deficient number, of specific

cell types causes the disease. Parkinson's disease, diabetes type I, spinal cord injury and stroke are just a few of a long list of diseases theoretically suited for stem cell based therapy.

Although hES cells have been available for more than eight years, not a single clinical trial has been reported started. However, a number of pre-clinical trials are ongoing, indicating that clinical trials are likely to be started in the near future.

1.3.4 Statement Regarding Future Scientific Development

It is impossible to predict the future of the stem cell field in detail. However, some general trends can be described. hES cells have demonstrated their usefulness as a tool for basic research, and provide a valuable supplement to multipotent stem cells from adult individuals. hES cells have furthermore emerged as a symbol of modern stem cell research, facilitating investment into the stem cell field. In relation to clinical usage hES cells face a number of challenges. These include the risk of immunological rejection, potential serious side effects like cancer, many ethical concerns, potentially very high treatment costs, and challenging technology transfer from laboratory to clinic. In addition, governmental approval may be cumbersome to obtain. These challenges apply only to a limited extent to somatic stem cells from adults. However, typically adult stem cells still have limitations, especially with regard to clinical use due to their rareness and difficulty of propagation in culture.

In summary, we believe that adult stem cells are more likely to become the treatment of choice for most patient groups if the isolation and the cell expansion challenges are satisfactorily resolved.

1.4 Philosophical and Theological Traditions

Within philosophical and theological debates certain main arguments have been used regarding the moral status of human embryos. We shall mention some of the argumentative concepts and evaluate their relevance, strengths and weaknesses.

1.4.1 Personhood

From a philosophical and theological standpoint, one has often referred to the fact of being a person, to 'personhood' or 'personality', as a criterion for deciding whether a human life has an unique value and right to life and protection. In some cases, one has identified 'personhood' with being human. In other cases, one has distinguished between being a person and being a human. The criterion of personhood has thus in practice not been very clarifying in the debate. In the first place, different spokespersons have used different criteria regarding the physical and/or psychological characteristics that mark a person. In the second place, opinions vary as to when these qualities are thought to be present in the development of human life. In the third place, the discussion of the relevance of the concept of person with a view to the value of unborn life has been imprecise in its use of biological, psychological, ethical, and legal language. In the fourth place, it is unclear whether the concept of person is an either-or concept, or a gradualist concept. The project group finds therefore that the distinction between person and non-person is not suited for the identification of the moral status of human embryos.

1.4.2 Potentiality

The potentiality argument states that that which has the potentiality of becoming a developed human being with moral status, has a right to life.

Two objections are often raised against this argument:

- First, the notion of potentiality is held to be unclear. Thus, for example, the further development of the embryo is dependent not only on the genetic potential, but also on other factors, such as for instance insertion into the uterus, the care of the mother and health professionals etc. However, stem cell research itself is based on the notion of potentiality. What makes stem cells so important is their potentiality to develop into any kind of cell in our body. And this development depends not only on the stem cell, but also on its being given the right kind of protective environment and growth conditions. Such factors do not contradict the embryo having its potentiality from the very beginning.
- Secondly, it has been objected to the potentiality argument that although a child is potentially an adult, an adult has many rights that the child does not have, for example the right to vote. However, the potentiality argument does not claim that the embryo has all the rights of a fully developed human being. Some rights, like the right to vote, or the right to practice medicine, require a certain age, a certain education, etc. The right not to be exposed to pain requires ability to feel pain, etc. But the right to life does not seem to require any such extra conditions, and it is difficult to see what can exclude an embryo from this right, given that it has potentiality of becoming a developed human being.

The group agrees upon the ethical relevance of this argument. But there are different views among us regarding the weight and the consequences of the argument.

Some in the group hold that ascribing human embryos potentiality does not necessarily include an absolute right to life and protection from harm, but an increasing right to life through pregnancy (see section 1.4.4).

Others maintain that such potentiality implies certain rights, such as right to life, right to care and to protection etc. They oppose the view that such rights emerge only later in the development.

1.4.3   Biological Continuity

The main substance of this argument is that there is a continuous development in the life of an embryo, beginning with fertilization, without any possibility of differentiating between clear stages with corresponding, variable right to life.

Against this it has been maintained that biological continuity does not exclude that there are morally relevant stages during the life of an embryo. According to this view, new elements are introduced, for instance when the primitive streak is created after approximately 14 days or when the brain is beginning to form. The development of an embryo includes initiation of new capacities throughout the pregnancy.

However, all members of the group agree that there are no morally relevant reasons for drawing sharp lines between different stages in the development of the embryo. Although we agree on the relevance of this argument, we differ regarding its strength.

Some members of our group evaluate this as a strong argument for full protection of a human embryo from its beginning through all phases. They see fertilization as the starting point of an uninterrupted biological development and a life history without any morally relevant leap with regard to protection of life. The sperm and the egg cells do not have this continuous identity with the fully developed human being. During its development, the embryo and later the fetus acquire new features, such as the ability to feel pain, and these features give it more rights, for example the right to be protected against pain, but there is no stage in the development where the right to life is changed.

Other members agree that there is a continuity in the biological development of the embryo, but do not consider this as a decisive argument for full protection of a human embryo from its beginning. This view is called the gradualist view. The gradual biological development yields a gradual increase in right to protection.

1.4.4.   Graduality

Graduality is an alternative for answering the question: At what point does a human being obtain its full moral status: At fertilization? At some point during pregnancy? At birth or at some point after birth? And how does it happen? As a complete either-or change? As a step by step process? Or as a gradual continuous process?

Some of us will argue from the full moral status of a grown up human being with its mental capacity, with a consciousness of itself, with rationality, with an understanding of the future etc. But does it follow that an embryo, which has these capacities only in potentiality, can still be ascribed the same moral standing? According to this position, this is not the case. The line of argumentation is such that we have to base our understanding of this on our moral feelings and intuitions, modified through a process towards a reflective equilibrium. We have to accept what such intuitions tell us: that the moral status of an embryo or a fetus is a growing process towards a full moral standing some time during pregnancy. This means that potentiality counts, but not by yielding a full right to life from fertilization. These members find that the rights of an embryo in its earliest phase, for example before organ and neuronal formation, in some cases can be weighed against other rights and highly valuable purposes.

Other members of the group argue that the right to protection of the embryo and the fetus does not depend on the degree of biological and psychological development. Our obligations towards human life in this phase do not presuppose that the embryo has acquired certain organs or consciousness. In addition these members maintain that our moral feelings and intuitions are not unambiguous and satisfactory sources for insight regarding ethical rights and duties.

1.4.5 Individuality

The argument from individuality states that a human embryo should be treated as an individual human life from the time when it is clear that 'twinning' is no more possible. This is supposed to be about 14 days after fertilization. According to this view it is up to this time not possible to securely identify an embryo as an individual human life.

It has been argued that this biological aspect is ethically relevant. Consequently, one will differentiate between a human embryo before 14 days and an individual human life after this limit, regarding the kind of protection which it deserves: During the first two weeks of its existence an embryo must be shown respect, but only an individual human life has a full right to be protected.

All members of the group hold that individualization has no relevance to the moral status of human life before and after the 14-day limit. The only difference may be numeric. A single embryo with the potentiality of twinning is still entitled to protection, even though it could split into two individuals.

1.4.6 God's Creation

In the stem cell debate religious arguments have also been used. Here we shall concentrate on a Christian perspective. Churches and theologians have often used what can be called a creation argument in their evaluations of acceptable ethical actions regarding the human embryo: A human being is a unique creation, brought into existence by a divine act and with the right to live and not be harmed. A human, considered in this manner, is presupposed to be a reality already from fertilization. This is sometimes supported with references to biblical texts that speak of human creatures as being created 'in the image of God' (Gen. 1:26–28, 5:1 f., and 9:6 f.), an expression of human uniqueness and of humanity's special role within creation, in contrast to all else that exists.

Within the group there is agreement that humans have uniqueness and a special role within the framework of existing reality. There is disagreement, however, concerning the value that can be ascribed to human life based on references to Christian belief in creation. Some wish to give this understanding universal ethical relevance as an articulation of an exclusive standard of value for exemplars of the human species. They consider this as possible regardless of whether one bases the valuation on Christian faith in God, in a non-Christian, religious position or in a non-religious conception of life. Others point to the existing disagreement within society with regard to the use of human embryos in research. They maintain that religious arguments for an ethical evaluation of humanity in all phases of its life can not be valid except as internal arguments within a fellowship of religious believers.

1.5   Normativity and Terminology

The project group is of the opinion that the most adequate approach to the question of the moral status of the embryo with a view to using embryonic stem cells in research and therapy, is found in an *anthropological and cultural approach*, where one identifies various concepts and categories that are connected to human uniqueness and expression, and decides which normative content these carry. In addition to the normative aspects in section 1.4, we will here focus in part on elements connected to human biological life, and in part on norms with relevance for research and clinical therapy.

1.5.1   Human Life

A concept such as *'human life'* can be given content from different perspectives. Biologically, 'life' can be defined based on the criteria (movement, growth, reproduction etc.) that all life allegedly fulfills. In our context, it is important to maintain that we are not talking of the responsibility of moral agents toward life in general, but toward human life. And we are using this concept in a descriptive, biological sense. In this statement human life is not understood as human material in general (eggs, cells etc.), but as the entity existing from the fertilization and onwards. According to this a human embryo is human life.

1.5.2   Human Being

The project group has been aware that probably no one will be able to deliver a complete and precise definition of what a human being is in terms of moral status. For the research project that involves biology/medicine, philosophy, and theology, the choice of professional perspective will be of great importance. A possible approach that can be agreed upon is to claim that a human being is a creature that descends from a woman and a man. In any case, one can maintain that a human being is an entity who biologically belongs to the species Homo sapiens.

In the opinion of the group, it is necessary to underline that an embryo originating from human beings is itself a human life, a *human* embryo. But there are different evaluations of whether such an embryo should be regarded as a *human being* or not. The reason for differing views is the normative implication of stating that a human *life* is a human *being*.

1.5.3 Dignity

Several normative concepts have been used in order to express the valuing of human life. From the side of an existing human life: sanctity, dignity, worth, value, inviolability etc. From the side of moral agents: reverence, respect etc. In this context we will concentrate on three of them: dignity, value and respect.

The idea of a unique dignity in an ethical sense for human beings is a central and fundamental part of both the humanistic and the Christian moral tradition. This dignity is held to be universal, understood as a common normative standing for all human beings.

The project group differs, however, regarding the question of the justification of and possible variations within dignity. Some wish to anchor dignity in empirical characteristics that distinguish humans (rationality, sense of identity, perhaps the potential of achieving these). An important distinction here is the difference between early life phases and later stages of fetal development. A consequence of such differentiation is variable human dignity (step by step or gradualistic), lower in an embryo than in a viable fetus or in a born human being. Others in the group wish to anchor human dignity in humanity's belonging to a specific biological species and/or in relation to God as its creator. With such an anchor, independent of varying, qualitative characteristics, it follows that dignity is valid in its own right.

Both positions in the group maintain that dignity is the basis for normative standards, in the sense of duty to protect human life. In the one case, this is seen as a gradualist duty, in the other case as a unitary and stabile duty.

1.5.4 Value of Human Life

The concept of value is normally used for positive values. With regard to humans one must differentiate between the value of biological existence and the value of a complete life history.

The project group understands the value of human life as a value in itself, an intrinsic value, arising from the existence of a human. Such a value is not merely instrumental and is not dependent on consciousness and abilities. Nor is it deduced from what a human being has of importance for other humans, not a result of the valuation of others. Only on these premises is it possible to maintain a value of human life which is universal in the sense of common for all.

The high ranking of the biological life of humans is also a consequence of the fact that it is a condition for being able to register and receive all other ethical values or goods.

1.5.5   Respect

Generally it may be said that valuable entities shall be respected in the sense that they shall not be destroyed. In cases of highly valuated things or creatures a great respect will usually be required.

In daily life the degree of respect toward human beings may vary according to situation. In the public debate on stem cell research one can notice various claims of respect maintained.

The project group agrees that human embryos shall be shown respect in attitudes, words and actions. This means in a concrete way that cells, tissues and organs from embryos shall not be used for every kind of purpose (for instance as animal food or cosmetics). But there are different understandings within the group with regard to the question of how radical or absolute this respect shall be applied. Some maintain that it is not contrary to this respect to use hES cells from embryos when there is an ethical, superior aim, such as new biological insights or the possibility of therapeutic progress. Others claim that respect for human life is ethically so fundamental and weighty that it can not be exempted from by referral to an eventual, new knowledge in research. They interpret the concept respect in this context as including the duty not to do harm.

1.5.6   Knowledge as Ethical Value

In addition to the anthropological elements already mentioned there are some important ethical norms or values linked to human, cultural activities such as research and clinical treatment.

Knowledge within natural sciences, social sciences and humanities is generally evaluated as being of great importance on both an individual and a communal level. Within the framework of ethical reflection we find it justified in a broad spectrum of different secular and religious ethical systems.

Medical knowledge is a main ethical value both with regard to its potential utility for an advanced health care system, but also for its basic insights before any application of it in relation to human illness and disorder.

1.5.7   Health

Health is a worldwide accepted, important ethical value with relevance for humans from fertilization until death. It is a central responsibility for each individual and for the society in common to keep and to restore one's own and other humans' health. This positive concern for health and physical care has to be combined with a continuous duty to eliminate or reduce illness, suffering and pain.

All kinds of work aiming at preserving human health may also be ethically justified by additional norms or principles like mercy, beneficence, justice etc.

1.5.8   Quality of Life

This concept has been used both descriptively and normatively in modern bioethics, but it has often been difficult to give it a sufficiently precise content in order to serve as a criterion for ethical decisions. Generally it seems to include two aspects: On the one hand, some 'objective' requirements are identified and must be met, like bodily functions, fulfilment of basic needs, social care etc. On the other hand, it is also a question of how an individual subjectively experiences one's own life situation. Quality of life must not be misused as a way of measuring people's capabilities or life style leading to a ranking of humans contrary to the fundamental idea of an equal human dignity.

An adequate understanding of the quality of life principle presupposes that it is interpreted as a concept expressing the purpose of a 'good life'. Having such a meaning, the value quality of life may be linked to other well-known ethical standards such as neighbourly love, community etc. It serves as a reminder of the duty to secure every citizen an acceptable level of life conditions beyond the biological existence itself.

1.6   Medical Progress and Respect for Embryos

1.6.1   Ethical Dilemmas

An ethical dilemma is usually defined as a conflict where some ethical norms support one solution of the actual case and other norms support another solution incompatible with the first. In the context of our research project the overall ethical dilemma is the conflict between on the one hand the ethical value of the embryo's human life and our respect related to this status, and on the other hand the possibilities of medical progress that can save lives and reduce suffering. There is no easy way out of the dilemma, no way to act according to all the ideal standards relevant in the challenging situation. We therefore in each situation have to weigh the sum of the norms and values counting for different alternative solutions.

Some within the group claim that before doing so, we have to take into account whether there are some borders that create a framework within which the final conclusion has to be drawn. They state that the inviolability of living, human existence is such a fundamental, ethical demand that it can only be exempted from it in some extreme situations. They understand this demand as relevant also regarding human embryos and do not evaluate the requirement for hES cells as an exceptional case. Consequently they support stem cell research using other alternatives than hES cells.

Others are convinced that a breakthrough in research for medical purposes is more likely with the use of hES cells. A main reason for this is their utility for getting new basic insights into the biological development of human beings and eventual disorders within this growing process. Additionally hES cells are, as already mentioned, pluripotent and able to differentiate in culture into all types of cells. They therefore find that using them in research is acceptable as a kind of balancing between respect for embryos over against the need to advance medical progress.

1.6.2 Purposes

Some of the purposes for stem cell research in general, including research on hES cells, correspond to essential ethical values mentioned above as health and quality of life. At the same time there are most likely also other purposes pushing individual researchers and research milieus forward: honour, career, economic profit etc. Stem cell research is expected to result in a range of new insights into biological processes within human development. We all agree in welcoming such knowledge.

There are also sufficient reasons for having expectations for the future regarding new kinds of clinical therapy. Some pictures of prospects are obviously examples of overselling the possible outcomes. We need to be aware of this danger and cautious in our hopes. Nevertheless, pointing to the usefulness of stem cell research is not necessarily a sort of improper utilitarianism, but an application of the raison d'être for all health systems: care for life, health and non-suffering.

On this point there is agreement among the members of the group.

1.6.3 Means

We agree that in scientific research of any kind it is not sufficient to evaluate the aims of research projects against central ethical values. It is also necessary to evaluate the means which are going to be used in each specific project.

In medical research there are already established international, ethical standards and codes dealing with the responsibility of researchers working with human beings. In several such documents special attention is drawn to research on children or on people without the ability to give their consent. An important ethical norm stated in such cases is the respect for the integrity and vulnerability of persons.

Some members of our group extend such restrictions also with regard to unborn human beings. Human integrity and vulnerability is never more important than when we are confronted with the weakest of all human life. They point to the danger of an instrumentalization of early human life which may violate the intrinsic value of such life.

Other members underline that they see research on hES cells not separated from, but within the framework of the whole field of such research. They accept that research on embryos is not unproblematic from an ethical point of view. But they do not find the reasons for avoiding all use of hES cells sufficiently weighty.

1.6.4   Alternative Sources

A number of alternative sources to the developing, human embryo have been proposed during the last few years (cf. section 1.3.2.2). These alternatives are commonly characterized by aiming at identifying less ethically problematic routes to embryonic stem cells. The group as a whole agrees upon a positive valuation of the intention of these alternatives and we will here briefly discuss some of these.

1.6.4.1   Viable Embryonic Stem Cell Lines Derived from 'Dead' Embryos

Embryos incapable of further embryonic development ('dead embryos') have been demonstrated to be a potential source for embryonic stem cell lines.

From a biological point of view, this alternative might be problematic since it is likely that the same causes that prevented the embryo from further development also might render the stem cells different from their normal counterparts. For example they may have various genetic abnormalities.

To prove, beyond doubt, that a given embryo is in itself incapable of further embryonic development requires markers. Such markers have, to date, not been established and it is likely that a number of normal embryos will be needed in research if such a marker is to be established.

A less rigid method of identifying embryos incapable of further development is to use an experienced embryologist to sort between healthy and unhealthy embryos. A high percentage of fertilized eggs do not develop naturally and will not be used in assisted reproduction. These embryos are in some laboratories being used successfully to derive embryonic stem cell lines.

The potentiality argument is important for all members of the group for distinguishing between ethically acceptable and unacceptable uses of embryos. Given that it can be proven that certain embryos do not have the potential to become a child, all members of the group find it ethically acceptable to use these embryos for research. However, the group will emphasize that it today is impossible to distinguish between embryos with and without the potential to develop to term and that the developing of such markers might be problematic since it might involve discarding healthy embryos.

1.6.4.2 Single Blastomeres Withdrawn from the Embryo Without Destruction

Based on experience with Preimplantation Genetic Diagnosis (PGD) we know that single cells can be withdrawn from developing embryos. Although not proven, the embryos do not seem to be harmed by the process. After withdrawal, in theory, the cell can be used to establish stem cell lines.

All members of the group consider this strategy for developing stem cell lines problematic. This is because it increases the risk on the developing embryo without providing the embryo (or the person when born) with advantages. An exception can for all members of the group be made if PGD is to be used, anyway, to exclude embryos with serious, inheritable diseases, and if the stem cell lines can be established without increasing the risk on the developing embryo. However, we recognize the difficulty in defining what are to be considered serious inheritable diseases and who shall make this decision.

As a consequence of the argument above, some members of the group also find it, in some rare cases, ethically acceptable to use embryos identified with a serious inheritable disease as a source for research. Others find that such a use is a kind of instrumentalization of seriously ill embryos.

1.6.4.3 Somatic Cell Nuclear Transfer (SCNT) and Altered Nuclear Transfer (ANT)

SCNT (also called therapeutic cloning) means the creation of an embryo by a fusion of a somatic cell nucleus and an enucleated, unfertilized egg. Some in the group find this alternative acceptable, given the possible scientific progress that may be achieved. Others reject it for the following reasons: (i) human embryos are created solely for research purposes, (ii) a high number of unfertilized eggs will need to be procured from fertile women, and (iii) successful development of SCNT might open the way to reproductive cloning of humans.

ANT (introduction of an altered nucleus) has been proposed as a means to obtain ethically unproblematic hES cells. The advantage of this alternative is that nothing that could become a human being is destroyed. The aim of the method is to create an unorganized cell mass, which lacks the potential to develop into a viable embryo. If this strategy can be implemented effectively for producing normal hES cells, it could be an important step towards ethically unproblematic hES cells. However, this alternative will require a high number of unfertilized eggs, it will need normal hES cell lines as controls and it will be difficult to prove, beyond doubt, that genetically altered embryonic cells do not have the potential to develop to term.

1.6.4.4   Redifferentiation of Somatic Cells

Redifferentiation of adult, somatic cells to a state enabling the derivation of embryonic stem cell lines is a promising, although very early, line of research. Recent research involving genetic manipulation has demonstrated that it might be possible to redifferentiate cells taken from adults to a state resembling embryonic cells.

Since no embryo is destroyed or harmed, this strategy is of little ethical concern to all members of the group. However, some members of the group find it problematic if normal embryonic stem cells will be used in this research as controls. Concerns can also be raised whether one is able to stop the redifferentiation at the stage where the most potent stem cells are pluripotent and not continue all the way to a totipotent cell enabling cloning.

1.6.5   Social and Political Context

The whole group emphasizes the great possibilities given through stem cell research for gaining new insights in embryological development and in the causality and the healing of serious diseases. At the same time it is important to practise a critical evaluation of the priority given to stem cell research, including research on hES cells, in our societies. This must be compared to the use of money and personnel in other fields of medical research, as well as in education, health system, development aid etc. In our context here it is necessary not only to deal with the moral status of human embryos, but also to focus on the moral status of medical research. It may be argued that other sectors of social and political engagement would save more human lives if given similar resources. In addition we have to take into account that therapeutic use of stem cell research may be expensive and in conflict with a wider, global responsibility for justice and equality in medical care. The challenge to promote health and quality of life by using new ways of treatment should not be limited to rich countries.

1.7   Conclusion

Among the members of the project group there is to a great extent agreement with regard to central aspects of the topic. We all evaluate the research going on in the stem cell field in general as supportable and promising for basic insights and potential future medical therapy. There is no disagreement in the group that from a biological and medical point of view hES cells are useful for such purposes. But we differ when it comes to the question of the acceptability of using embryos as sources. Further we agree upon giving ethical relevance to traditional arguments such as potentiality and continuity in favour of the moral status of human embryos. We also share the understanding of an embryo as being a human life. And we all find that dignity, value of human life and respect are central ethical standards with regard to our responsibility towards embryos. Medical research and clinical treatments are ethically justified by values such as knowledge, health and quality of life. And finally, we stand together in supporting research on alternative sources to human embryos and in underlining the importance of both serious, ethical evaluations of individual projects in the stem cell field, and of public debates about the priority of such research in society.

Summarizing, we nevertheless find that there are two different positions in our group regarding the main problem of the project:

- The first position emphasizes that hES cells are useful for basic and applied research and in the future potentially also for treatment of currently untreatable, serious diseases. The supporters take as their starting point the biological fact that the development of a human embryo is a continuous process which progresses through several phases, and that the moral status grows accordingly up to full status when the fetus has completed its development during pregnancy. In consequence, they interpret the substance of norms like dignity, value of life and respect as growing regarding the responsibility of moral agents towards human embryos. They find that a reference to potentiality does not necessarily include an absolute right to life and protection.
- The second position includes a positive evaluation of the research on human stem cells, except the use of hES cells derived from human embryos. The group members holding this opinion maintain that somatic stem cells have certain advantages compared to hES cells with regard to prospects for clinical therapy. They also point to the possibility of using alternative sources for hES cells. As a starting point for their ethical argumentation they focus on human dignity as a stable, uniform norm including right to life and integrity also for embryos. They interpret the value of human life as an intrinsic norm of high rank and consequently view respect in relation to such a life as fundamental. This group argues that the potentiality of the embryo implies a right to life, care and protection. And they see continuity as a strong argument for the embryo's uninterrupted life history without any morally relevant jump, and reject a valuing based upon phase of development or certain qualities. They understand a reference to creation as a supplement to a more general reference to the uniqueness of humans.

Part I
Biological and Medical Perspectives

Chapter 2
Stem Cells: Sources and Clinical Applications

Steinar Funderud

Abstract Research within the field of stem cells has especially during the last two decades given incredibly much new insight into how organs and tissues in our bodies undergo replenishment or repair. In this short review some of most recent developments within somatic stem cells and embryonic stem cells are briefly discussed with the intention to serve as a background for the ethical discussions in later chapter of this book. It is important to bear in mind that the stem cell field is still in its infancy where we in despite of the overwhelming amount of data already existing, will require much more research to establish safe and reliable therapeutic protocols.

Keywords Somatic stem cells, embryonic stem cells, stem cell potency

2.1 Introduction

The idea of tissue regeneration is not a new observation of today, but goes back to an ancient Greek myth where Prometheus according to the myth transgressed the law of the gods when he gave the fire to mankind. As punishment for this act he was chained to a rock, and an eagle was sent each day to eat his liver. However, during the nights the liver regenerated and Prometheus survived. While it has been known for several decades that lower animals like amphibians can regenerate an amputated limb, it is only recently that stem cell research has brought data disclosing that this ancient myth in fact is a reality for most tissues of the human body including liver.

Although the former proof of the stem cell concept in mammals came almost 50 years ago by the work of McCulloch and Till (Till and McCulloch 1961), which demonstrated that different blood cell lineages originated from a common stem

---

Ullernchauseen 70, Montebello, NO-0310 Oslo, Norway.
e-mail: steinar.funderud@rr-research.no

cell, the full implication of stem cells in medicine has come only recently with numerous reports on isolation of stem cells from different tissues of the human body and not at least the successful derivation of human embryonic stem cell lines (Thomson et al. 1998). These achievements have led to great push towards development of technologies for culturing and differentiation of stem cells ex vivo for the purpose of tissue regeneration.

The scope of this chapter is not to give a comprehensive review of the stem cell field as such, but rather to briefly review topics serving as a background for the ethical discussion bearing mainly on the source of stem cells applied in the field of stem cell research.

2.2   Stem Cell Definition and Potency

Stem cells are undifferentiated, self-renewing cells with the potential to produce specialized differentiated cells. The potency of a given stem cell is reflecting the purpose the stem cell is serving in the tissue it derives from.

Stem cells derived from the inner mass cells of the blastocyst 5–10 days after fertilization of the egg are called embryonic stem cells (ES), and resident stem cells in born individual are named somatic stem cells (SSC). SSC are also referred to as adult stem cells because they derive from adult tissue.

Stem cell potency is referred to as the potential to give rise to a range of new cell phenotypes. There are three basic measures of stem potency: Totipotent, pluripotent and multipotent stem cells. A fertilized egg is the ultimate stem cell as it contains the potency to differentiate into all cells of the three embryonic germ layers and extraembryonic cell types. Fertilized eggs are accordingly a totipotent cell. ES cells are pluripotent and can differentiate into all cell types of the three germ layers, but they have lost the ability to differentiate into cells of the extraembryonic tissue. Most SSC cell types are multipotent and produce usually cells restricted to a related family of cells, e.g. hematopoietic stem cells which differentiate into cells of the different blood cell lineages. Unipotent stem cells are restricted to production of one differentiated cell type only.

2.3   Somatic Stem Cells

For centuries scientist have known that different species of lower animals can regenerate missing parts of their bodies. Mammals have not the same capacity to regenerate body parts, but have evolved systems to maintain function of the different tissues through activation of stem cell pools harbored in the tissue of question. In adult mammals most tissues exhibit low turnover under normal circumstances, however, some tissues such as skin, gut, blood and testis constantly renewing. This observation was for some time interpreted as existence of stem cell function in some tissues and lack of stem cells in others. However, tissues thought of as not renewing like, e.g. brain and heart, have during the last decade been shown to exhibit cell turnover and renewal from somatic stem cells although at a lower rate. Today the common perception is that all organs harbor stem cells. The majority of SSC are multipotent and essential for homeostasis of most tissues, organ and cell systems throughout the lifespan of mammals. Multipotent stem cells are currently considered to be tissue specific in vivo, but there might be exceptions to this where we in future might manipulate a certain stem in vitro to serve other purposes than in vivo.

2.3.1 Stem Cells in Blood and Bone Marrow

The two major multipotent stem cell types in the bone marrow are hematopoietic stem cells (HSC) and mesenchmal stem cells (MSC). The HSC is a well characterized multipotent stem cell type. Their main location is in specific niches in the red bone marrow. The niches function both as a sanctuary and stimulatory site where the stem cells can get signals for both proliferation and differentiation into the different blood cell lineages at a rate of about 10^{11} to 10^{12} new blood cells every day. HSC is also found in the blood and in umbilical cords, however, here they do not proliferate. HSC can relatively easily be isolated from sources like bone marrow, blood and umbilical cords, and have been applied clinically for several decades already.

Mesenchymal stem cells are another important stem cell residing in bone marrow. MSC have a wide differentiation repertoire and are in vivo differentiating into various cells in tissues like bone, cartilage, tendon, adipose tissue, bone marrow stroma and smooth muscle cells. MSC are well characterized and easily obtainable from bone marrow or adipose tissue. Efficient protocols for expansion and differentiation in vitro are available. Studies in mice suggest that MSC may have potential for diverse clinical applications. Thus several clinical protocols like: repair of damaged myocardium in ischemic patients, prevent graft versus-host-disease, repair of bone in children with osteogenesis imperfecta among others have been initiated.

2.3.2 Stem Cells in Skin and Intestine

Dermal and intestinal epithelium is object to environmental insults and exhibit accordingly high cell turnover which must be replenished instantly. Skin is for instance in humans renewing itself every two weeks. The current understanding is that there arc several types of skin stem cells. Dermal stem cells found in the epidermis, melanocyte stem cells and epidermal stem cells. Epidermal stem cells are residing in a structure of the hair follicle named bulge, which is the stem cell niche for the epidermal stem cell. From the bulge they can differentiate into specialized cell of the hair follicle and the interfollical epidermis.

Replenishment of epithelial lineages of the gastrointestinal tract is as frequent as skin occurring every two to seven days. This process is regulated by stem cells in stem cell niches of the intestinal crypts. From here the stem cells differentiate into the different cell types of the intestinal epithelium.

2.3.3   Stem Cells in the Brain

Neural stem cells (NSC) are besides blood stem cells the most studied stem cell type in humans, however, they are still not as well characterized as blood stem cells. NSC appears to be localized in two major regions in adult brains. One population resides in the subventricular zone, and the other population in dentate gyrus of the hippocampus. Some reports from studies of rodent brains suggest that stem cell derived progenitors migrate from the subventricular zone into the olfactory bulb suggesting a role in regeneration, however, no such data is yet available for the human brain. Numerous studies have demonstrated that human NSC can be isolated from adult brain tissue and expanded in vitro. Moreover, such cells may be induced to differentiate into all three major neuronal cell types, including neurons, oligodendrocytes and astrocytes. Such cells may be applied in future therapies of brain disorders, e.g. Parkinson's disease and degenerative diseases of the central nerve system. (For more details see chapters 4 and 5, this volume.)

2.3.4   Stem Cells in the Heart

Multipotent cardiac stem cells have just recently been identified and localized in adult hearts. Such stem cells were originally observed in developing heart tissue. Thus it was doubt about the existence of intrinsic stem cell based repair in the adult heart. Now it is well established that stem cells exist in such tissue although at a lower number than young individuals. Cardiac stem cells are reported to possess the ability to differentiate into myocytes, endothelial cells and smooth muscle cells which constitute the major cell types of the myocardium. However, as for the brain, there is still a poor understanding of the role of heart stem cells in the repair of damaged heart tissues in vivo.

2.3.5   Stem Cells in the Pancreas

With the growing number of persons with diabetes comes the question of the existence and role of pancreatic stem cell in the homeostasis of beta-cells. Results from transplantation of beta-cells from cadaver donors have encouraged the search for pancreatic stem cells. Much evidence indicates the presence pancreatic stem cells in ductal or islets regions, but the localization of the stem cell niches is not precisely known. Pancreatic stem cells/progenitor cells have recently been isolated from pancreatic duct-epithelium, and such cells have been proved to be capable of differentiating into endocrine cells. However, a cure based on such cells is still far ahead.

2.3.6 Stem Cells in the Eye

Three different compartments of stem cells have been in identified in various regions of the eye; cornea, the conjunctiva and retina. Of these three is the stem cells of the cornea, limbal epithelial stem cells, in routine clinical use. The existence conjunctival stem cells is well documented, but the precise location in the conjunctiva is not well defined. The detection of retinal stem cells has led to studies on such cells in restoration of a functional retina in, e.g. patients with age-related blindness like macular degeneration.

2.3.7 Stem Cells in the Lung

Stem populations have been described in different tissues of the lung like the upper lung airway epithelium, in alveoli and in bronchioles. The role of lung stem cells in homeostasis of lung tissue has for some time been controversial, but now it is well established that the different stem cell populations affects both daily turnover and repair of lung injury.

2.3.8 Stem Cells in the Ear

A population of stem cells residing in the adult inner ear has recently been discovered. These stem cells are capable of differentiating into the inner ear hair cells, and thus raise the hope for a future development of a stem cell based treatment.

2.3.9 Stem Cells in the Liver

Liver has an impressing regeneration capacity, and can within weeks regenerate up to 50 percent of its mass strongly suggesting the existence of potential liver stem cells. Oval cells were for some time incorrectly referred to as the liver stem cell, but the currant understanding is that oval cells are the progeny of a liver stem cell, and not a stem cell itself. It behaves like bipotential progenitor cells capable to differentiate into mature hepatocytes and bilial epithelial cells. Oval cells share some phenotypic characteristics of hematopoietic stem cells which led to the idea that hematopoietic stem cells might be useful for liver regeneration. This has proved to be an erroneous idea. The recent understanding is that hematopoietic stem cells are not capable of reconstituting damaged liver tissue.

2.3.10   Stem Cells in the Skeletal Muscle

Injured skeletal muscle may be repaired from different stem- or progenitor cells residing in the muscle tissue. One of these cells is satellite cells located at the periphery of skeletal myofibers. Satellite cells are dormant cells which can be triggered to proliferate for both self renewal and differentiation into myogenic cells.

2.4   Embryonic Stem Cells

Pluripotent stem cells can be derived from mainly three different sources. The most applied source for derivation of pluripotent stem cells are left over eggs from the infertilization clinics. Another source for pluripotent stem cells is testicular teratocarcinomas from where embryonic carcinoma (EC) cells can be isolated. A very interesting third possibility was recently suggested by a research group in Japan (Takahashi and Yamanaka 2006). By combining four selected transcription factors crucial for maintaining pluripotency, they were able to generate pluripotent cells from adult fibroblast cultures. Such pluripotent stem cells are named iPS.

2.4.1   hES Cells

The first successful derivation of human embryonic stem cell (hES) by Thomson and coworkers in 1998 relied on the gradual process made in the previous almost 20 years of work with murine embryonic stem cells. Later several laboratories around the world have developed hES cell lines applying the same technology, and today more than hundred different well characterized cell lines are available for research. One decisive parameter in the original culture protocol for hES cells is the presence of feeder cells, and until very recently hES had to be maintained in cultures with mouse fibroblasts feeder cells to keep the ES cells in an un-differentiated stage. Unfortunately, such culture conditions are a prohibition for future clinical application, as cell lines deriving from such cultures may contain retroviruses of murine origin. This problem has now been solved through extensive search for alternative culturing conditions. Thus, several laboratories have been successful in developing stable hES lines which have not been in contact with any xenogenic products.

The capacity of hES cells to differentiate into different cell types of all three germ layers of the human body highlights their promising role in regenerative medicine. There is an instant increase in reports on successful derivation of applicable cell types, and clinical protocols are underway. Geron, a cell therapy company in California, has recently got permission from US Food and Drug Administration (FDA) to start a clinical phase I trial in patients with spinal cord injury in 2007. In this study the patients will be receiving oligodendrocyte precursors. The long term goal for the study is to apply such cells in restoration of the myelin sheath which is normally lost during spinal cord injuries. The present clinical trial is a safety study with no expectation of cure for the patient included. However, studies in rats strongly indicate that therapeutic cells injected within a short limit of time of injury lead to an improved functionality in the rats. Although this is promising data it is anticipated that spinal cord injured patients are likely to need additional injection of astrocytes for replacement of lost neurons for complete recovery of the injury. Experts in the field foresee at least another 5–10 years research before spinal cord patients can be helped from such therapies.

2.4.2   ES Stem Cell Research

The ES stem cell research has indeed advanced rapidly in the last few years. Several recent studies in animal models have reported promising data applying different tissue specific cells derived from hES cells. When transplanted into animal disease models, hES derived cells have proved to be capable of restoring a variety of injured tissue. In a rat model hES derived oligodendrocyte progenitor cells improved motor function through remyelation, in a primate model dopaminergic neurons promoted behavioral recovery of Parkinson's disease, in a pig model cardiomyocytes restored ischemic hearts, and in a mouse model pancreatic islet cells reversed hyperglycemia in diabetic mice.

Although ES cells have a large growth and differentiation potential, there are several limitations and questions linked to ES cells in clinical use. One major issue is that undifferentiated hES cells may lead to teratomas in a clinical setting, because undifferentiated ES cells are basically a tumor forming cell. Another major problem facing widespread use of ES cells in cell therapy, is the anticipated rejection of transplanted cells by the patient's immune system due to tissue mismatching. There are of today only a limited number of hES cell lines available, which make it very unlikely to find a cell line matching the tissue type of a given patient. Transplantation of cells deriving from lines with poor tissue matching will potentially cause graft failure. Therefore, strategies to avoid or overcome graft rejection have to be devised. One obvious opportunity is to develop banks of hES cells with diverse tissue types represented. A recent data simulation in UK estimated that a bank of about 150 different hES lines might be sufficient for providing a satisfactory match for 25–50 percent of potential recipients, and 95 percent chance for a full match for 8 percent of the patients. Given that this is a correct estimate it is manageable to build such a bank, however, there will still be only a minority that may be helped from such banks. Another long term possibility to cope with the rejection issue, is to induce immune tolerance to the donor cell in the recipient. This concept has so far been demonstrated to work with ES cells in a rat model. However, a broader applicability and the underlying mechanisms involved are yet to be established.

2.4.3 Somatic Cell Nuclear Transfer (SCNT)

The possibility of graft rejection is a serious obstacle to a wide clinical applicability of hES technology. This problem is circumvented by applying genomic replacement technology made famous through the cloning of the sheep Dolly by Ian Wilmut and coworkers. In this technology the nucleus of a somatic cell is transferred to an egg devoid of its own genetic material. The cytoplasm of the egg harbor factors which rejuvenilate the somatic cell nucleus, and the embryonic developmental program is started. At the stage of the blastocyst ES cells isogenic to the donor of the nucleus may be derived. The success of nuclear transfer of Dolly fueled the stem cell debate around the world, because the fear of a cloned human. The advocates for the technology invented the term therapeutic cloning because they had no intention what so ever to clone humans. Most countries around the world took of moral reasons stand against the technology, and US like most European countries have still a ban on it. Some countries, e.g. like Great Britain, Sweden and Australia has lifted the ban and allows nuclear transfer under certain restrictions. The correct acronym for somatic cell nuclear transfer is today SCNT. Several laboratories around the world are trying to work out protocols for SCNT in humans, however, we are still waiting for the first successful report.

Aside technical problems with SCNT in humans, which most likely will be solved in near future, SCNT still hinges on both ethical and medical issues linked to the use of human unfertilized eggs. Another major concern to opponents of hES technology is that ES derivation from blastocysts involve embryo destruction. To get around this obstacle different alternative methods are suggested. One method, advocated by William Hurlbut, is based on the observation that mouse embryos carrying a mutation in the Cdx2 gene die at the blastocyst stage because the outer layer of the blastocyst is not formed. The cells of the inner mass can still give rise to ES cells. Hurlbut proposes that hES cells can be derived in an equivalent process in humans knocking out the human CDX2 gene. The method is called alternative nuclear transfer (ANT). (For more details see chapter 6)

2.4.4 Induced Pluripotency (iPS)

Recently another interesting method for derivation of pluripotent stem cells has been suggested by Yamanaka and coworkers (Takahashi and Yamanaka 2006).

The idea behind their work was to reprogram somatic cells to pluripotency by introducing key transcription factors functioning in the maintenance of pluripotency in embryonic cells. They were able to generate stem cells from mouse fibroblasts and called the stem cells induced pluripotent stem (iPS) cells. Transplantation of such cells into nude mice resulted in tissues from all germ layers. If iPS can be accomplished in humans scientists should be able to develop stem lines from patients which suffer from different genetic diseases. Such lines would be invaluable research tools for understanding specific diseases.

2.5 Prospects and Controversies in the Stem Cell Field

A tremendous amount of reports on stem cells, especially in the last decade, have raised the prospect of regenerating most failing organs or tissues of the body from such cells. Much excitement originates from laboratory studies, or studies in mouse models with the expectations that such results would easily be translated into cures in the clinic. Especially reports on the plasticity of SSC created hopes for clinical applications in different patient groups which will be difficult to satisfy. This turned out to be a too hasty conclusion as most of these studies were solely in vitro studies. There are several reasons for the erroneous conclusions. One reason was that a limited number of phenotypical markers acquired under artificial conditions in vitro were taken as proof for transdifferentiation capacity in some SSC. Change in some surface markers does not necessarily mean changed functional capability. Another mistake came from animal studies which indicated that stem cells from one tissue could replace cells in another tissue. It turned out that the injected stem cells had fused with resident cells and not changed phenotype and function. The question of transdifferentiation is still controversial, but the dominant perception today is that this is not a likely phenomenon which can be applied in regenerative medicine, although only future studies can answer this question with certainty.

Tissues which actively regenerate from resident SSC represent the lowest hurdle for tissue regeneration. Such tissue has appropriate stem cell niches that provide the correct factors for proliferation and differentiation of stem cells. When therapeutic stem cells are homing to such tissue they will start to proliferate and differentiate into functional cells. The success of bone-marrow transplantation underline the viability of this principle. However, what we can expect from stem cells in tissues or organs which do not undergo an extensive regeneration like blood and skin is not known at present.

There are other examples of successful clinical applications of SSC than bone marrow transplantation, e.g. eye stem cells in repair of cornea, however, for tissues like brain, heart and pancreas which have a low regeneration rate it is still a lot more to learn before we see routine clinical protocols. Generally there are two major obstacles for SSC in regenerative medicine. Low accessibility of stem cells from the tissue in question, and lack of expansion protocols ex vivo. Thus, this has to be worked out before SSC can be safely applied in regenerative medicine. This is hopefully within reach in a few years time at least for some patient groups.

For ES cells the situation is different. In contrast to SSC has ES cells got a growth and differentiation potential which make them ideal for production of large quantities of therapeutic cells. They are pluripotent, meaning that they can differentiate into any cell type of the body. Consequently ES has the capacity to be a renewable cell source for a multitude of cell based therapies. Thus the challenge is to translate the capacity of ES cells into safe clinical protocols. Such a protocol has already been worked out for the neural linage, and as briefly outlined above has Geron got a FDA approval on a phase I study where spinal cord injured patients will receive oligodendrocytes derived from hES cells. Other protocols heading at specialized cell types like beta-cells of the pancreas and heart myocytes are under way. ES cells therefore present itself as the ideal source for regenerative medicine, however, as dwelt with above there are several severe issues connected to ES cells in tissue regeneration: the ethical issue linked to derivation of ES cells, tissue type matching and the safety issue linked to a possible tumor hazard in vivo. The problem connected to the ethical and the matching tissue type issue may find its solution in the suggested alternative routes for creation of pluripotent cells. The safety issue is also within reach through a proper testing of cell batches before transplantation.

When comparing hES cells and SSC in the prospect of future clinical application it is important to keep in mind that ES research is still in its infancy, and that it will take many years to mature into safe clinical protocols. We do still have a lot to learn about the mechanisms in tissue regeneration in general. Thus to move forward it is vital that both adult and embryonic stem cell research is pursued collaboratively in parallel. The goal of the future is to supply the clinic with phenotypically and functionally defined stem cells or stem cell derivatives.

References

Takahashi, K. and Yamanaka, S. (2006). "Induction of pluripotent stem cells from mouse embryonic and adult fibroblast cultures by defined factors", *Cell* 126, 663–676.

Thomson, J.A., Itskovitz-Eldor, J., Shapiro, S.S., Waknitz, M.A., Swiergiel, J.J., Marshall, V.S., and Jones, J.M. (1998). "Embryonic stem cell lines from human blastocysts", *Science* 282, 1145–1147.

Till, J.E. and McCulloch, E.A. (1961). "A direct measurement of the radiation sensitivity of normal mouse bone marrow cells", *Radiation Research* 14, 213–222.

Chapter 3
Alternative Means to Obtain Pluripotent Stem Cells

Ole Johan Borge

Abstract The isolation and use of pluripotent stem cell lines from human embryos is ethically controversial because it normally involves the destruction of embryos. Pluripotent human stem cells are by definition able to produce all cell types in adult individuals. Thus, they are highly valuable as research tools for both basic and applied research, in addition to an anticipated usage in regenerative medicine. Currently, the human embryo is the main source of pluripotent stem cells. Recently, however, a number of alternative strategies has been proposed indicating that pluripotent stem cells can be obtained without destructing viable human embryos. I will here give an introduction to some of these alternative strategies.

Keywords Pluripotent stem cells, alternative sources, dead embryos, parthenogenesis, ANT

The announcement of the derivation of pluripotent stem cell lines from human blastocysts (Thomson et al. 1998) and aborted fetuses (Shamblott et al. 1998) in November 1998 placed stem cells at the center stage of medical research and marked the starting point of a large and emotional bioethical debate. Pluripotent stem cells were immediately seen as modern medicines Holy Grail by holding the promise to treat every disease caused by lack of any given cell type. The interests of suffering patients and the freedom of research were set against the moral status of the human embryo.

Supernumerary embryos after assisted reproduction and elective aborted fetuses were the only sources of pluripotent stem cells that were debated in the first few years after 1998. The search for alternative, less ethically challenging, sources reached general public attention firstly when the US President's Council on Bioethics published its report on *Alternative sources of human pluripotent stem*

The Norwegian Biotechnology Advisory Board, Pb. 522 – Centrum, N-0105 Oslo, Norway
e-mail: ojb@bion.no

L. Østnor (ed.), *Stem Cells, Human Embryos and Ethics: Interdisciplinary Perspectives.*   31
© Springer Science + Business Media B.V. 2008

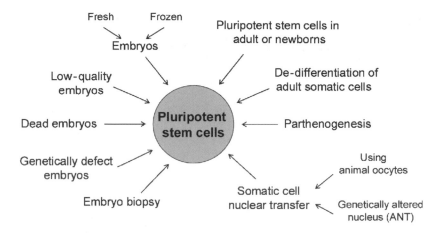

Fig. 3.1 The figure depicts some alternative routes to obtain pluripotent stem cells. Some of these routes are well documented. Others, have however only been demonstrated in animal models or to a limited degree in humans

cells in 2005. The search for alternatives has been driven by the desire to find less ethically problematic sources and to obtain pluripotent cell lines eligible for federal funding in the US. Currently, only 21 embryonic stem cell lines are available to federally founded research in the US (NIH 2007).

I will below briefly, describe some of the sources that can be used to isolate pluripotent stem cells (Fig. 3.1).

3.1 Supernumerary Embryos

Human pluripotent stem cells are typically isolated from embryos 5–6 days after in vitro fertilization. At this stage the embryo has formed a hollow sphere of cells, called a blastocyst, containing approximately 100 cells. The blastocyst has an outer layer of cells and an inner cell mass. A few cells within the inner cell mass will normally develop into the fetus, whereas the rest will form extraembryonic tissues, like the placenta, needed for fetal development.

Supernumerary embryos can either be fresh, viable embryos not needed for assisted reproduction or frozen embryos that have expired the legal storage time. In most countries there is no alternative use of supernumerary embryos than to discard them or to use them for research. In a few countries, however, it is allowed to put embryos up for adoption. Although embryo adoption is allowed, it will not significantly reduce the high number of embryos discarded every day, world wide.

Embryos can also be made specifically for research purposes. This can be done either in conjunction with harvesting egg and sperm cells for assisted reproduction or germ cell donation solely for research.

3.2 'Dead' Embryos

During assisted reproduction an average of 10 unfertilized eggs are harvested after a hormone treatment of the women. After mixing egg and sperm, typically three eggs will not be fertilized; another three stop at various stages of early embryonic development or demonstrate abnormal development; whereas four develop nicely and meet the quality requirements for implantation to achieve pregnancy. The three embryos that stop during development are most likely incapable of further embryonic development and unable to develop to term if implanted. However, some of the developmentally arrested embryos do contain viable cells and these can potentially be used to isolate pluripotent stem cells (Landry and Zucker 2004). It has recently been demonstrated that pluripotent stem cells isolated from developmentally arrested embryos are comparable to pluripotent stem cell lines isolated from healthy embryos (Zhang et al. 2006). This work did however reveal that only embryos that arrested late (6–7 days after fertilization) could be used to derive embryonic stem cell lines. Embryos that were arrested early (3–5 days after fertilization) did not generate stem cell lines.

An important question is how to ensure that any given embryo is indeed 'dead' and incapable of further development. Should this simply be decided by visual examination, or do we need rigid physiological or biochemical markers? If so, such markers have not been identified to date. Furthermore, if only late stage arrested embryos can be used it can be argued that this is unethical unless we also can ensure that the in vitro culture conditions are equally efficient in nurturing the embryo as the microenvironment in the uterus.

3.3 Genetically Defect Embryos

Preimplantation genetic diagnosis (PGD) is used to identify embryos with genetic disease. This is normally done by withdrawal of a single cell from three-day-old embryos. The severity of the genetic diseases possible to diagnose using PGD range from diseases with a certain death during pregnancy to less severe conditions that might occur many years into adult life. Commonly, embryos diagnosed with genetic disease are not implanted and simply discarded.

Alternatively, these embryos could be used as sources for pluripotent stem cells (Verlinsky et al. 2005). It can be envisaged that pluripotent stem cell lines with a known genetic condition will be an attractive research tool. Furthermore, it can be argued that stem cell lines isolated from embryos carrying a genetic mutation, not compatible with life, would be of little ethical controversy.

3.4 Single Blastomers

PGD has demonstrated that a single cell can be withdrawn from early embryos seemingly without harming the remaining cells. Instead of using the single blastomers for genetic analysis, they can be used to derive embryonic stem cell lines.

If successful, this strategy makes it possible to obtain both pluripotent stem cells and a living human being from one and the same embryo. This has been demonstrated in mice (Chung et al. 2006; Wakayama et al. 2007) and to some degree in humans (Klimanskaya et al. 2006; Vogel 2006).

Whether this is an ethical unproblematic alternative depends in part on whether the withdrawal of the single blastomer increases the risk of destroying the developing embryo, or if human embryonic stem cell lines are still needed to nurture the single blastomer during the first few days in culture.

3.5   Somatic Cell Nuclear Transfer (SCNT)

Embryos can in theory be generated by different cloning techniques. Embryo cloning has been demonstrated in animals by removing the nucleus of an unfertilized egg and thereafter replacing the removed nucleus with the nucleus of a cell taken from the animal to be cloned (Kawase et al. 2000; Munsie et al. 2000; Wilmut et al. 1997). The egg containing the transferred nucleus is then stimulated to divide until it reaches the blastocyst stage. Thereafter the cells of the inner cell mass can be isolated and pluripotent stem cells derived.

Cloning involves the use of a high number of unfertilized eggs. Human eggs are very hard to obtain (Cyranoski 2007) and will most likely include paying the egg donor a significant amount of money (Daley et al. 2007). To avoid the problem of procuring human eggs, it has been proposed to use animal eggs from slaughter houses or laboratories instead. This has even been tried, and reported to be successful by using eggs from rabbits (Chen et al. 2003). Recently, another promising strategy for obtaining enucleated human eggs for research was proposed (Egli et al. 2007). In brief; eggs fertilized with two sperm cells are incapable of embryonic development, but instead of simply discarding these eggs, they might be useful in research (Colman and Burley 2007).

Somatic cell nuclear transfer using enucleated animal eggs can be viewed as very ethically problematic, since it involves the mixing of two species. Although the nucleus is removed the animal egg will still provide mitochondria to the developing embryo. However, until high numbers of mature human eggs can be produced in vitro, the use of animal eggs can potentially be a practical solution if we want to make somatic cell nuclear transfer workable in the foreseeable future.

Altered somatic cell nuclear transfer (ANT) is a variant of cloning where the transferred nucleus is altered before transfer to the enucleated egg (Hurlbut 2005; The President's Council on Bioethics 2005). The goal with this strategy is to generate normal pluripotent stem cells but without at any time or stage of development having an embryo with the capability to develop into a fetus. This has been tried, with success, in mice (Meissner and Jaenisch 2006).

Although this strategy is proposed solely to produce ethically unproblematic human pluripotent stem cell lines much remains to prove this point. For example this strategy requires a high number of unfertilized human eggs, and furthermore it will be difficult to prove beyond doubt that the generated cells are indeed incapable of embryonic development if not truly tested by transfer into the uterus. Even further, although philosophical consistent, a priory it does not appear ethically sound to generate temporarily genetically crippled embryos only to satisfy those opposing research on human embryos (Holden and Vogel 2004).

3.6 Parthenogenesis

Unfertilized eggs can be triggered to divide, without fertilization – this is called parthenogenesis. Although parthenogenesis results in viable offspring in some insect species, mammalian parthenotes die during the first few days of embryonic development (Kiessling 2005). However, using laboratory tricks, a few mice have in fact been born after parthenogenesis (Kono et al. 2004) and embryonic stem cell lines have been obtained from embryo-like structures generated after parthenogenesis in both mice (Kim et al. 2007) rabbits (Wang ct al. 2007), monkeys (Vrana et al. 2003), buffaloes (Sritanaudomchai et al. 2007) and humans (Revazova et al. 2007).

Taken that no embryo is destructed and that the unfertilized eggs are obtained in ethically acceptable ways, this strategy is of little ethical concern. However, as mentioned, it is currently very difficult to obtain human eggs for research.

Furthermore, it remains to verify that embryonic stem cell lines from parthenotes are equally stable, efficient and malleable as cell lines isolated from normally developed embryos.

3.7 De-Differentiation of Specialized Cells from Adults

The differentiation from a stem cell to a specialized cell has been believed to be a one-way-route without the possibility to differentiate back to a more undifferentiated state. However, the birth of cloned animals has made it clear that the DNA present in specialized cells do contain all the genetic material necessary for making an entire new individual (Wilmut et al. 1997). Thus, making embryonic stem cells from differentiated cells should, in theory, be possible.

Researchers are underway in identifying genes and corresponding proteins needed for reprogramming specialized cells back to an embryonic stage. This research is being carried out by fusing cells and infusing genes, proteins or simply cell extracts, known to be important for pluripotent cells (Collas and Hakelien 2003; Cowan et al. 2005). Surprisingly, published results indicated that a rather low number of genes might be responsible in defining a given cell type and differentiation stage (Takahashi and Yamanaka 2006). This work was recently refined by demonstrating that only four genes could de-differentiate murine skin cells to cells similar to embryonic stem cells (Maherali et al. 2007; Okita et al. 2007; Wernig et al. 2007). This was made possible by advanced gene manipulation and, although very promising, this is still a very early field of research (Rossant 2007).

Since no embryo is destroyed or harmed, this strategy is of little ethical concern. However, some might even find this alternative problematic since human embryonic stem cells are likely to be used in this research as controls. Concerns can also be raised whether one is able to stop the de-differentiation at a stage where the most primitive cell is pluripotent and not de-differentiate all the way to a totipotent cell and thereby enabling cloning.

3.8   Pluripotent Stem Cells in Newborns and Adults

Although pluripotent stem cells exist both in embryos and fetuses, it is presently not clarified whether such cells also exist in newborns, children, adults or tissue normally discarded at birth. However, several reports during the last few years indicate that pluripotent stem cells may exist even in adult individuals. Pluripotent-like cells have been isolated both from the bone marrow (Jiang et al. 2002) and testis (Cyranoski 2006; Guan et al. 2006) from adults. Other results even indicate that pluripotent-like stem cells might exist in the amniotic fluid (De et al. 2007), cord blood (Kogler et al. 2004) and placenta (Miki and Strom 2006) at the time of birth.

In general, these various pluripotent-like cells have not yet been thoroughly tested to state, without doubt, that they are comparable to embryonic stem cells or that they might be used as interchangeable alternatives in research or medical treatment.

Generally, there are few ethical concerns if cells can be isolated from materials that otherwise will be discarded at birth, without harming the newborn, or from renewable sources from adults able to give an informed consent.

3.9   The Alternative Sources and the Future Development

As briefly described above a number of different strategies can, at least in theory, be used to isolate pluripotent or pluripotent-like stem cells. But what are the minimum criteria these alternative cell types must meet to be considered as ethically unproblematic and scientifically comparable to embryonic stem cell lines?

Many oppose to research that involves the destruction of viable embryos. Others are even against all in vitro fertilization that might potentially lead to supernumerary embryos. Another line of ethical objections could be raised toward strategies increasing the risk of the embryo without at the same time providing essential benefits to the coming child. Furthermore, it can be argued that it is unethical to ask women to donate unfertilized eggs as long as it involves hormone treatment or economical incentives to donate.

To use animal instead of human eggs is fiercely debated and though some find it very ethical problematic others are more pragmatic and argue that this is a practical way to avoid the ethical and practical obstacle by procuring human eggs.

Others might argue that strategies that are proposed simply to circumvent national regulations and that lack a strong research rationale, should be avoided unless they generate pluripotent stem cells of equal or superior quality. In particular, those that find the isolation of pluripotent stem cells from supernumerary embryos acceptable, might furthermore see this entire debate merely as a distraction of focus and an academic exercise without the likelihood of resolving any controversy in the stem cell field.

I think that the search for alternative ways to obtain pluripotent stem cells will continue with increased interest. There are several reasons for this. Firstly, the limited federal funding in the US and restrictive legislation in several countries attract scientists to find less ethically problematic sources. Secondly, therapeutic cloning has yet to demonstrate its potential in providing patient-specific pluripotent stem cells and thus, the hunt for transplantable stem cells will continue. Finally, the significant commercial opportunity available is likely to attract both academic and corporate research.

In conclusion, I hope that this brief presentation about alternative strategies to obtain pluripotent stem cells will encourage scientists and others to look deeper into this matter and that the debate about ethically acceptable sources will continue as the field of stem cell research evolves in the years to come.

References

Chen, Y., He, Z. X., Liu, A., Wang, K., Mao, W. W., Chu, J. X., Lu, Y., Fang, Z. F., Shi, Y. T., Yang, Q. Z., Chen, da. Y., Wang, M. K., Li, J. S., Huang, S. L., Kong, X. Y., Shi, Y. Z., Wang, Z. Q., Xia, J. H., Long, Z. G., Xue, Z. G., Ding, W. X., and Sheng, H. Z. (2003). "Embryonic stem cells generated by nuclear transfer of human somatic nuclei into rabbit oocytes", *Cell Res.*, vol. 13, no. 4, pp. 251–263.

Chung, Y., Klimanskaya, I., Becker, S., Marh, J., Lu, S. J., Johnson, J., Meisner, L., and Lanza, R. (2006). "Embryonic and extraembryonic stem cell lines derived from single mouse blastomeres", *Nature*, vol. 439, no. 7073, pp. 216–219.

Collas, P. and Hakelien, A. M. (2003). "Teaching cells new tricks", *Trends Biotechnol.*, vol. 21, no. 8, pp. 354–361.

Colman, A. and Burley, J. (2007). "Stem cells: recycling the abnormal", *Nature*, vol. 447, no. 7145, pp. 649–650.

Cowan, C. A., Atienza, J., Melton, D. A., and Eggan, K. (2005). "Nuclear reprogramming of somatic cells after fusion with human embryonic stem cells", *Science*, vol. 309, no. 5739, pp. 1369–1373.

Cyranoski, D. (2006). "Stem cells from testes: could it work?", *Nature*, vol. 440, no. 7084, pp. 586–587.

Cyranoski, D. (2007). "Teams trail genes for human 'stemness' ", *Nat. Med.*, vol. 13, no. 7, p. 766.

Daley, G. Q., Ahrlund, R. L., Auerbach, J. M., Benvenisty, N., Charo, R. A., Chen, G., Deng, H. K., Goldstein, L. S., Hudson, K. L., Hyun, I., Junn, S. C., Love, J., Lee, E. H., McLaren, A., Mummery, C. L., Nakatsuji, N., Racowsky, C., Rooke, H., Rossant, J., Scholer, H. R., Solbakk, J. H.,

Taylor, P., Trounson, A. O., Weissman, I. L., Wilmut, I., Yu, J., and Zoloth, L. (2007). "Ethics. The ISSCR guidelines for human embryonic stem cell research", *Science*, vol. 315, no. 5812, pp. 603–604.

De, C. P., Bartsch, G., Jr., Siddiqui, M. M., Xu, T., Santos, C. C., Perin, L., Mostoslavsky, G., Serre, A. C., Snyder, E. Y., Yoo, J. J., Furth, M. E., Soker, S., and Atala, A. (2007). "Isolation of amniotic stem cell lines with potential for therapy", *Nat. Biotechnol.*, vol. 25, no. 1, pp. 100–106.

Egli, D., Rosains, J., Birkhoff, G., and Eggan, K. (2007). "Developmental reprogramming after chromosome transfer into mitotic mouse zygotes", *Nature*, vol. 447, no. 7145, pp. 679–685.

Guan, K., Nayernia, K., Maier, L. S., Wagner, S., Dressel, R., Lee, J. H., Nolte, J., Wolf, F., Li, M., Engel, W., and Hasenfuss, G. (2006). "Pluripotency of spermatogonial stem cells from adult mouse testis", *Nature*, vol. 440, no. 7088, pp. 1199–1203.

Holden, C. and Vogel, G. (2004). "Cell biology. A technical fix for an ethical bind?", *Science*, vol. 306, no. 5705, pp. 2174–2176.

Hurlbut, W. B. (2005). "Altered nuclear transfer: a way forward for embryonic stem cell research", *Stem Cell Rev.*, vol. 1, no. 4, pp. 293–300.

Jiang, Y., Jahagirdar, B. N., Reinhardt, R. L., Schwartz, R. E., Keene, C. D., Ortiz-Gonzalez, X. R., Reyes, M., Lenvik, T., Lund, T., Blackstad, M., Du, J., Aldrich, S., Lisberg, A., Low, W. C., Largaespada, D. A., and Verfaillie, C. M. (2002). "Pluripotency of mesenchymal stem cells derived from adult marrow", *Nature*, vol. 418, no. 6893, pp. 41–49.

Kawase, E., Yamazaki, Y., Yagi, T., Yanagimachi, R., and Pedersen, R. A. (2000). "Mouse embryonic stem (ES) cell lines established from neuronal cell-derived cloned blastocysts", *Genesis*, vol. 28, no. 3–4, pp. 156–163.

Kiessling, A. A. (2005). "Eggs alone", *Nature*, vol. 434, no. 7030, p. 145.

Kim, K., Lerou, P., Yabuuchi, A., Lengerke, C., Ng, K., West, J., Kirby, A., Daly, M. J., and Daley, G. Q. (2007). "Histocompatible embryonic stem cells by parthenogenesis", *Science*, vol. 315, no. 5811, pp. 482–486.

Klimanskaya, I., Chung, Y., Becker, S., Lu, S. J., and Lanza, R. (2006). "Human embryonic stem cell lines derived from single blastomeres", *Nature*, vol. 444, no. 7118, pp. 481–485.

Kogler, G., Sensken, S., Airey, J. A., Trapp, T., Muschen, M., Feldhahn, N., Liedtke, S., Sorg, R. V., Fischer, J., Rosenbaum, C., Greschat, S., Knipper, A., Bender, J., Degistirici, O., Gao, J., Caplan, A. I., Colletti, E. J., meida-Porada, G., Muller, H. W., Zanjani, E., and Wernet, P. (2004). "A new human somatic stem cell from placental cord blood with intrinsic pluripotent differentiation potential", *J. Exp. Med.*, vol. 200, no. 2, pp. 123–135.

Kono, T., Obata, Y., Wu, Q., Niwa, K., Ono, Y., Yamamoto, Y., Park, E. S., Seo, J. S., and Ogawa, H. (2004). "Birth of parthenogenetic mice that can develop to adulthood", *Nature*, vol. 428, no. 6985, pp. 860–864.

Landry, D. W. and Zucker, H. A. (2004). "Embryonic death and the creation of human embryonic stem cells", *J. Clin. Invest.*, vol. 114, no. 9, pp. 1184–1186.

Maherali, N., Sridharan, R., Xie, W., Utikal, J., Eminli, S., Arnold, K., Stadtfeld, M., Yachechko, R., Tchieu, J., Jaenisch, R., Plath, K., and Hochedlinger, K. (2007). "Directly reprogrammed fibroblasts show global epigenetic remodeling and widespread tissue contribution", *Cell Stem Cell*, vol. 1, no. 1, pp. 55–70.

Meissner, A. and Jaenisch, R. (2006). "Generation of nuclear transfer-derived pluripotent ES cells from cloned Cdx2-deficient blastocysts", *Nature*, vol. 439, no. 7073, pp. 212–215.

Miki, T. and Strom, S. C. (2006). "Amnion-derived pluripotent/multipotent stem cells", *Stem Cell Rev.*, vol. 2, no. 2, pp. 133–142.

Munsie, M. J., Michalska, A. E., O'Brien, C. M., Trounson, A. O., Pera, M. F., and Mountford, P. S. (2000). "Isolation of pluripotent embryonic stem cells from reprogrammed adult mouse somatic cell nuclei", *Curr. Biol.*, vol. 10, no. 16, pp. 989–992.

NIH (2007). *Information on Eligibility Criteria for Federal Funding of Research on Human Embryonic Stem Cells.*

Okita, K., Ichisaka, T., and Yamanaka, S. (2007). "Generation of germline-competent induced pluripotent stem cells", *Nature*, vol. 448, no. 7151, pp. 313 317.

Revazova, E. S., Turovets, N. A., Kochetkova, O. D., Kindarova, L. B., Kuzmichev, L. N., Janus, J. D., and Pryzhkova, M. V. (2007). "Patient-specific stem cell lines derived from human parthenogenetic blastocysts", *Cloning and Stem Cells*. vol. 9 no. 3, pp. 432–449.

Rossant, J. (2007). "Stem cells: the magic brew", *Nature*, vol. 448, no. 7151, pp. 260–262.

Shamblott, M. J., Axelman, J., Wang, S., Bugg, E. M., Littlefield, J. W., Donovan, P. J., Blumenthal, P. D., Huggins, G. R., and Gearhart, J. D. (1998). "Derivation of pluripotent stem cells from cultured human primordial germ cells" [published erratum appears in Proc Natl Acad Sci U S A 1999 Feb 2; 96(3):1162], *Proc. Natl. Acad. Sci. U S A*, vol. 95, no. 23, pp. 13726–13731.

Sritanaudomchai, H., Pavasuthipaisit, K., Kitiyanant, Y., Kupradinun, P., Mitalipov, S., and Kusamran, T. (2007). "Characterization and multilineage differentiation of embryonic stem cells derived from a buffalo parthenogenetic embryo", *Mol. Reprod. Dev.*, vol. 74, no. 10, pp. 1295–1302.

Takahashi, K. and Yamanaka, S. (2006). "Induction of pluripotent stem cells from mouse embryonic and adult fibroblast cultures by defined factors", *Cell*, vol. 126, no. 4, pp. 663–676.

The President's Council on Bioethics 2005, *White paper: alternative sources of human pluripotent stem cells*.

Thomson, J. A., Itskovitz-Eldor, J., Shapiro, S. S., Waknitz, M. A., Swiergiel, J. J., Marshall, V. S., and Jones, J. M. (1998). "Embryonic stem cell lines derived from human blastocysts" [published erratum appears in Science 1998 Dec 4; 282(5395):1827], *Science*, vol. 282, no. 5391, pp. 1145–1147.

Verlinsky, Y., Strelchenko, N., Kukharenko, V., Rechitsky, S., Verlinsky, O., Galat, V., and Kuliev, A. (2005). "Human embryonic stem cell lines with genetic disorders", *Reprod. Biomed. Online.*, vol. 10, no. 1, pp. 105–110.

Vogel, G. (2006). "Stem cells. Scientists derive line from single embryo cell", *Science*, vol. 313, no. 5790, p. 1031.

Vrana, K. E., Hipp, J. D., Goss, A. M., McCool, B. A., Riddle, D. R., Walker, S. J., Wettstein, P. J., Studer, L. P., Tabar, V., Cunniff, K., Chapman, K., Vilner, L., West, M. D., Grant, K. A., and Cibelli, J. B. (2003). "Nonhuman primate parthenogenetic stem cells", *Proc. Natl. Acad. Sci. U.S.A*, vol. 100 Suppl 1, pp. 11911–11916.

Wakayama, S., Hikichi, T., Suetsugu, R., Sakaide, Y., Bui, H. T., Mizutani, E., and Wakayama, T. (2007). "Efficient establishment of mouse embryonic stem cell lines from single blastomeres and polar bodies", *Stem Cells*, vol. 25, no. 4, pp. 986–993.

Wang, S., Tang, X., Niu, Y., Chen, H., Li, B., Li, T., Zhang, X., Hu, Z., Zhou, Q., and Ji, W. (2007). "Generation and characterization of rabbit embryonic stem cells", *Stem Cells*, vol. 25, no. 2, pp. 481–489.

Wernig, M., Meissner, A., Foreman, R., Brambrink, T., Ku, M., Hochedlinger, K., Bernstein, B. E., and Jaenisch, R. (2007). "In vitro reprogramming of fibroblasts into a pluripotent ES-cell-like state", *Nature*, vol. 448, no. 7151, pp. 318–324.

Wilmut, I., Schnieke, A. E., McWhir, J., Kind, A. J., and Campbell, K. H. (1997). "Viable offspring derived from fetal and adult mammalian cells" [published erratum appears in Nature 1997 Mar 13; 386(6621):200], *Nature*, vol. 385, no. 6619, pp. 810–813.

Zhang, X., Stojkovic, P., Przyborski, S., Cooke, M., Armstrong, L., Lako, M., and Stojkovic, M. (2006). "Derivation of human embryonic stem cells from developing and arrested embryos", *Stem Cells*, vol. 24, no. 12, pp. 2669–2676.

Chapter 4
Neurogenesis and Potential Use of Stem Cells from Adult Human Brain

Håvard Ølstørn, Morten C. Moe, Mercy Varghese, and Iver A. Langmoen*

Abstract Neural stem cells are present in the adult human brain of mammals, including humans, and can give rise to the three major cell types of the central nervous system; neurons, astrocytes and oligodendrocytes. These stem cells hold great promise for neural repair after injury or disease, either by activating the stem cells residing within the brain and/or by transplantation of stem cells from the adult human brain after expanding them in culture dishes. Autologous transplantation, in which a patient is transplanted with cells derived from his or her own brain, could circumvent some of the problems associated with the use of embryonic stem cells or fetal tissue, in particular the ethical concerns and problems with immune rejection. However, it must be demonstrated that the necessary types of neural cells can be generated in sufficient amounts, and that they can induce long-lasting functional improvements in animal models of brain disease and injury.

Keywords Adult, human, neurogenesis, neural stem cells, transplantation

4.1 Introduction

Adult stem cells are undifferentiated cells found among mature and specialized cells in a tissue or organ. They reside in various tissues in the human body, with bone marrow, peripheral blood, skin, skeletal muscle and liver being well known examples. Other terms proposed for these cells are tissue stem cells or somatic stem cells. They can differentiate to yield the specialized cells of the tissue or organ, and their main function is to maintain and repair the tissue in which they are found (Pessina and Gribaldo 2006) (National Institutes of Health, NIH, web site on stem cell information: http://stemcells.nih.gov/info).

---

*Vilhelm Magnus Center/Department of Neurosurgery, Ullevål University Hospital 0407 Oslo, Norway. e-mail: laiv@uus.no

L. Østnor (ed.), *Stem Cells, Human Embryos and Ethics: Interdisciplinary Perspectives.*   41
© Springer Science + Business Media B.V. 2008

The adult human brain is an extremely complex organ, with more than 100 billion interconnected cells in networks and signaling pathways. The brain used to be viewed as static, in the sense that no new neurons were generated after birth. This dogma goes back to the early 1900s and the Spanish neuronanatomist and Nobel Prize laureate Santiago Ramon y Cajal, who stated that 'nothing may regenerate in the brain or central nervous system, everything may die' (Ramon y Cajal 1913). With new methods it has been established that neurogenesis, i.e. generation of new neurons, indeed occurs in certain regions of the mammalian brain, including humans (Eriksson et al. 1998; Johansson et al. 1999; Kukekov et al. 1999; Roy et al. 2000; Arsenijevic et al. 2001). Stem cells have two defining properties – they can self-renew to produce more stem cells and they can differentiate to generate specialized cell types (Goh et al. 2003). The stem cells found in the adult brain are multipotent, i.e. they can give rise to the major cell types in the central nervous system; neurons, astrocytes and oligodendrocytes (Kornblum 2007). Due to limitations in self-renewal and restrictions in differentiation, these cells are often referred to as progenitor or precursor cells (Emsley et al. 2005) Could these neural stem cells or progenitors in the adult human brain have a role in the treatment of disorders in the central nervous system, and as such represent an alternative to embryonic and fetal cell sources?

4.2   The Complexity of the Human Brain

In order to understand the challenges facing cell replacement therapies of brain disease or injury, it is necessary to give a very brief overview of the complexity of the brain.

The billions of cells in the adult human brain belong to two broad, but different categories. Firstly, neurons, or nerve cells, represent the main functional unit of the nervous system. They produce electrical signals and communicate with other neurons through tiny contact points, known as synapses. In the synapse, the electrical activity of the signaling neuron is translated into a chemical signal that modulates the activity of the target neuron.

The second cell type in the brain is glia. This class of cells consists of non-neuronal cells that provide support and nutrition and participate in signal transmission in the nervous system. Two main cell types of glia are the oligodendrocyte and the astrocyte. The main function of the oligodendrocyte is to produce a layer of fat, called myelin, around the nerve fibers, known as axons, which neurons send out to communicate with other parts of the nervous system. The purpose of this layer of fat is to provide electrical insulation of the nerve fibers and thereby increase the speed at which impulses travel along the fiber. The astrocyte was originally seen as a cell that only glued the functional units together, hence the name glia, but has later been found to also have other important roles in the brain, like metabolic support and regulation of the external chemical environment. In addition to this, the brain also contains other cell types, for instance cells associated with the blood vessels. These are not cells of genuine neural origin, as they originate from immature cells in the bone marrow, and invade the central nervous system during development.

Neurons can be further divided into many different subtypes, based for example on their electrical and chemical properties. One main type is the glutamatergic neuron, which releases the amino acid glutamate at its nerve endings (synapses) and excites the target neuron. These neurons often project over considerable distances in the central nervous system, for instance from the cerebral cortex down to the spinal cord, where they directly activate motor neurons that in turn trigger muscular contractions. Another common type has inhibitory actions, and is called GABAergic. This type releases the neurotransmitter GABA from its nerve endings and thereby inhibits activity in the target neuron. They are often so-called interneurons, involved in connecting neurons in specific regions within the brain. A third type of neurons is modulatory, in which several classes of neurotransmitters regulate diverse populations of neurons. Examples of such neuromodulators include dopamine, serotonin, acetylcholine and histamine.

4.2.1 Brain Diseases and Injuries

Most injuries or diseases causing changes in motor or sensory function, verbal communication, cognitive or emotional function, are caused by changes in the activity of the neuronal networks of the brain. Diseases and injuries come with different levels of complexity. On the one hand we may experience damage to relatively large regions of the brain in head injuries or when one of the larger blood vessels is blocked, as in stroke. In such cases there is not only loss of a very large number of cells, but also of a large number of different subtypes of neurons and the very complex connectivity between these cells.

On the other hand, other conditions are simpler, in the sense that cells of a particular type are affected. These include for instance Parkinson's disease, which is due to a specific loss of so-called dopaminergic cells, i.e. neurons that secrete dopamine at their nerve endings. These cells are located in a specific region of the midbrain known as the *substantia nigra* and send nerve fibers to the basal ganglia, an area deep in the brain that has a number of important functions, among them regulation of motor activity, enabling us to perform smooth and controlled movements.

A common denominator in the above-mentioned examples is the loss of neurons.

4.3 Tissue Regeneration

Many tissues in the body have considerable capacity for regeneration. We have all observed that small injuries to the skin heal spontaneously and perfectly. Somewhat larger injuries heal spontaneously with small scars and even quite extensive injuries heal well with simple aids like wound closure with sutures. In some animals, the regenerative potential is in many cases even more impressive. Newts can regenerate their tails and limbs, including the neurons that occupy these regions (Brockes 1997) and lizards can regenerate entire brain parts (Lopez-Garcia et al. 1992). Neurogenesis in the adult has been extensively studied also in animals with a nervous system closer to mammals in complexity, like in songbirds, as well as in mammals, especially rodents.

4.3.1 Regeneration and Neurogenesis in the Brain

In contrast to many other tissues in the human body, the central nervous system has a very poor regenerative capacity. The aforementioned axiom of the early neuroanatomists, stating that the adult human brain is unable to produce new neurons, was hard to argue against. It was challenged in the 1960s, but the results were not accepted by the scientific community (Emsley et al. 2005; Kornblum 2007). The challenge came from experimental work by the American neuroscientist Joseph Altman, who suggested a neuro-regenerative potential in certain brain areas (Altman and Das 1965; Altman 1969). Altman made small lesions in the brain of rats, followed by injection of radioactive thymidine, which is one of the four bases of DNA. The radioactive thymidine was used as an indicator of cell division, as it was incorporated in the DNA of new cells. When Altman later examined the rat brains, he observed cells that (1) looked like neurons, and (2) contained radioactive thymidine, the implication being that new neurons had been produced in the adult rat brain.

In biology it is a common notion that the strength of proof required is proportional to the originality of the hypothesis proposed. Regarding Altman's work, several objections could be raised. These included for instance that labeling of neighboring cells could give a false impression of neurons containing radioactive thymidine, and that he could not be sure that the labeled cells were neurons, as the proposed neurons were identified by morphological criteria alone. In addition to this came the common conception that one expected to see division of the neurons themselves if neuronal multiplication took place. His observations did therefore not make it to the textbooks and were almost forgotten.

It was not until the end of the 20th century evidence started to build up against the dogma. This began with observations made in the brain of songbirds. In the canary brain, for instance, the song system was found to consist of several separate, but interconnected centers. A number of these were small in female birds, which have a small musical repertoire, but large in male birds, which have complex, learned singing. This led to the idea that sex hormones may influence singing and the size of brain song centers, and experiments where females were given testosterone not only showed that these females started to sing more, but also that song nuclei in their brain could double in size. Moreover, the song centers in the brain of males showed seasonal variations; they were large during the breeding season when singing is instrumental in courting and decreased in size when the breeding season was over.

Nottebohm and his colleagues used radioactive thymidine to investigate if the enlargement of song centers was caused by an increased number of cells. Making very thin brain sections, much thinner than a single cell, they could exclude the possibility that positive staining came from neighboring cells (Nottebohm and Arnold 1976; Nottebohm 1981; Nottebohm 1985)..

Despite strong indications of new neurons in the song nuclei in birds, they did not see any signs of neuronal recruitment from the local cell population. By examining the whole brain, however, they discovered cell divisions in the ventricular wall and cell migration from this area to the vocal centers. They concluded that prior to breeding season new neurons are produced in the vocal centers by cells migrating from the ventricular walls.

More recently, new technical developments made it possible to reexamine the possibility of neuronal regeneration in mammals. To understand the importance of these techniques, keep in mind that in order to prove that new neurons are formed in the adult brain it is necessary to verify that the cell (1) was produced by a cell division occurring in adult life, and (2) that it actually is a neuron.

The first method that enabled new investigations was the so-called bromodeoxyuridine (BrdU) technique. BrdU is, like thymidine, integrated in the DNA of dividing cells. In addition it may be stained by specific antibodies attached to fluorescent molecules emitting light with a certain wave-length following excitation. Thus, if BrdU is administered to a living animal during the course of one day, all cells formed by cell division during that day will have incorporated BrdU into their DNA. Furthermore, if a BrdU-specific antibody emitting red light is applied to tissue sections made from this animal, then all cells containing BrdU will appear red under the fluorescent microscope.

The second method was the discovery of cell-specific markers, which is based on the fact that different cell types express different types of complex molecules or antigens that may be targeted by specific antibodies. Neurons, for instance, express a set of markers that are thought only to occur in neurons. Thus, simply put, if you to a tissue section apply a neuron-specific antibody attached to a molecule that emit green light then all green cells in that section should be neurons.

The third method was confocal laser microscopy. This microscope has superior resolution and can be used to examine histological sections in stepwise depth, thus ensuring that the marker you see is in the cell you are studying and not in a cell immediately below or above it.

In 1992, Reynolds and Weiss isolated cells from a part of the brain called striatum of adult mice and induced proliferation by epidermal growth factor (Reynolds and Weiss 1992). Subsequently subsets of the cells developed the morphology and antigenic properties of neurons and astrocytes. Some of the newly generated cells also expressed immunoreactivity for the neurotransmitters typically found in that area of the adult mouse brain. This was followed by a number of studies indicating that neurogenesis could take place in vivo in adult animals, and that populations of stem or progenitor cells giving rise to new neurons on a more regular basis could be found in the ventricular wall (lining the fluid-filled spaces in the brain) and in the dentate area (a part of the hippocampus in the brain, an area dealing with short term memory, learning and spatial orientation) (Gage 2000). There are also several studies suggesting that multipotent progenitor cells may exist in other regions of the central nervous system, but that the local environment is not permissive for neurogenesis (Gage 2000; Taupin and Gage 2002).

It took several more years, however, before it was proved that the new cells actually were neurons. In this context it is important to realize that a functional neuron only can be identified by proving that the cell in question has the following properties: Firstly, it must have the ability to generate the typical electrical signal of a neuron (short lasting, low threshold action potentials), and secondly it must have the ability to communicate with other neurons by synapses. Identification of neurons by immunohistochemical techniques alone is somewhat uncertain; we and other groups have seen multiple examples of cells expressing mature neuronal markers without the ability to produce mature electrical signals. In 2002, however, Fred Gage and collaborators published two studies where they had identified new-born cells in the adult rodent brain which had the capability of producing short lasting, low threshold action potentials and had established synaptic connections (Song et al. 2002; van Praag et al. 2002).

4.3.2   Regeneration in the Adult Human Brain

Until the late 1990s, it was quite unclear whether neurogenesis occurred in the adult human brain, as it for obvious reasons was unthinkable to administer research substances to living humans for the purpose of identifying newborn cells. BrdU is, however, sometime given to cancer patients for diagnostic purposes and Ericsson and co-workers used this unique opportunity (Eriksson et al. 1998). They were granted ethical permission to examine these brains post mortem and co-stained for BrdU and the frequently used neuronal marker NeuN. By doing that they could identify cells that both (1) were born during the period the patients had been given BrdU, and (2) expressed neuronal antigens. Other studies, utilizing brain tissue removed for instance during temporal lobe resections due to epilepsy, showed that it was possible to develop monoclonal cultures of cells from the adult human brain and differentiate these cells into mature cells with immunhistochemical characteristics of oligodendrocytes, astrocytes, and neurons, the three principal building blocks of the brain (Johansson et al. 1999; Kukekov et al. 1999; Arsenijevic et al. 2001; Nunes et al. 2003). Again, the question remained whether the new cells were functional, electrical neurons.

We wanted to investigate this essential question, and utilized the stem cell culturing techniques and the immunohistochemical techniques developed by our collaborators at the Karolinska Institute in Stockholm, together with electrophysiological recordings (patch-clamping techniques) in our laboratory at the University of Oslo (Westerlund et al. 2003). In this study we observed that over a period of three weeks, neuron-like cells gradually developed electrical features, ending with quite mature-looking action potentials. Thus, we were not only able to reproduce earlier findings of cells derived from the adult human brain developing into oligodendrocytes, astrocytes, and neurons, based on the antigens they expressed. More importantly, we showed that the glia-like and neuron-like cells had functional properties according to their specific type.

Later on, we found cells expressing both neuronal markers and specific markers of synapses (Moe et al. 2005b). These results suggested that a number of the cells developed into neurons. It was, however, still necessary to demonstrate that the cells developed functional synaptic connections and more mature electrical signals.

We therefore decided to prolong the period of differentiation and were able to keep the cells for at least four weeks. The extensive epilepsy program at Rikshospitalet in Oslo allowed sufficient tissue for systematic patch-clamp studies at different stages of development. We found that over a period of four weeks in a culture dish, the cells went through characteristic steps of morphological and electrophysiological development in a similar manner to new neurons formed in vertebrates at the embryonic stage (Moe et al. 2005a, b).

Regarding the synapses, and the crucial question of communication with other newly formed neurons, we also combined immunohistochemical and electrophysiological techniques. We found expression of both glutamate receptors and GABA receptors, indicating that the cells also had the capability to communicate by chemical transmission. As mentioned earlier, the electrical signals traveling along a nerve fiber is transformed into a chemical signal in the synapses. Again, the functional evidence for synaptic connections came from patch-clamping of individual cells, and we found spontaneous synaptic events mediated by the release of both glutamate and GABA, similar to the action occurring in synapses in vivo (Hablitz and Langmoen 1982; Moe et al. 2005b). Furthermore, we studied pairs of neuron-like cells simultaneously, to find out whether the cells actually communicated with each other, i.e. that a signal in one cell led to a response in the next. We identified a number of cell pairs where action potentials in one cell generated a synaptic current in the other cell, thus directly demonstrating that the cells had developed functional synaptic contacts (Moe et al. 2005b).

4.4   Strategies for Neuronal Replacement

The fact that the adult human brain contains stem cells and progenitors capable of generating new neurons, astrocytes and oligodendrocytes brings hope of novel neural repair strategies in the future (Lie et al. 2004; Emsley et al. 2005; Galvin and Jones 2006; Pessina and Gribaldo 2006; Kornblum 2007).

The idea of cell replacement as a treatment strategy in certain neurological disorders is not new and actually preceded the discovery of stem cells in the adult mammalian brain. By the early 1980s it had become clear that the type of neurons that degenerate in patients suffering from Parkinson's disease has certain chemical similarities to cells in the medulla of the adrenal gland. Experiments in rats with modeled Parkinson's disease showed that transplantation of adrenal medullary cells could relieve some of the symptoms. Following this, Backlund and his colleagues transplanted tissue fragments from the adrenal medulla into the brains of a few patients with severe Parkinson's disease and observed some rewarding, but transient, effects (Backlund et al. 1985; Backlund 1987).

Today, several different strategies are being explored, and almost all studies are preclinical, using laboratory animals. Many models for brain lesions and diseases are in use, like neurodegenerative disease (for example models for Parkinson's disease [Svendsen et al. 1997; Ostenfeld et al. 2000; Liker et al. 2003]), focal ischemia or stroke (Arvidsson et al. 2002; Kelly et al. 2004; Thored et al. 2005), global ischemia (Nakatomi et al. 2002; Olstorn et al. 2007), brain trauma (Riess et al. 2002; Wennersten et al. 2004) and even models of multifocal disease like multiple sclerosis (Pluchino et al. 2003).

In principal, there are two main strategies of repair, namely stimulation of endogenous stem cells and transplantation of stem cells or fetal tissue.

4.4.1 Endogenous Repair in the Adult Brain

One potential source of neural stem cells for neural repair is through the mobilization of endogenous stem cells in the brain (Lie et al. 2004; Emsley et al. 2005; Kornblum 2007). This strategy is illustrated in the landmark study by Nakatomi and colleagues, who infused growth factors into the brain of adult rats following the selective degeneration of neurons in a certain part of the hippocampus called CA1, caused by global ischemia (when the blood flow to the rat brain is stopped transiently during surgery) (Nakatomi et al. 2002). The growth factors stimulated the neural stem cells in the rat brain and a significant number of new neurons were found, which regenerated and seemed to replace many of the neurons lost in the CA1. The new neurons survived at least up to six months after the injury, and importantly, the authors found evidence suggesting that the neurons were integrated into the local neuronal circuitry. Furthermore, behavioral studies showed that the growth-factor treated animals had better recovery and performed better in tests of spatial orientation and memory. Other studies have also showed an endogenous response towards lesions in the brain, an effect that can be relatively long-lasting (Arvidsson et al. 2002; Thored et al. 2005). However, the functionality of the new neurons created after injury, and the causality between new neurons and behavioral improvement, is not completely understood (Lie et al. 2004; Thored et al. 2005). Furthermore, as always with animal models, we do not know how findings in these models will translate into the human system (Lie et al. 2004). Still, with increased knowledge of the biology of endogenous progenitors, neuronal replacement strategies based on their manipulation may be possible in the future (Emsley et al. 2005).

4.4.2 Transplantation to the Adult Brain

The other main strategy for cell therapy is to transplant cells into the brain. The cells or tissue for transplantation studies may come from different sources, like fetal and adult rodent neural stem cells, fetal and adult human neural stem cells, neural cells derived from embryonic stem cells and also non-neuronal cells, for instance derived from the bone marrow (Mezey et al. 2000; Kornblum 2007; Olstorn et al. 2007).

Transplantation studies may also differ with respect to what is 'expected' of the cells used. Actual replacement of dying or lost neurons requires a lot from the grafted cells; in addition to robust survival, the cells need to fully integrate into the complex pathways in the adult host brain and take over the function of the lost cells (Svendsen and Caldwell 2000; Kempermann et al. 2004). This, however, might not be necessary in all cases to achieve a therapeutic effect. Transplanted cells also may have bystander effects, like inducing self-repair and protecting brain cells at a site of tissue damage by releasing growth factors or other trophic agents(Tai and Svendsen 2004; Martino and Pluchino 2006).

4.4.2.1 Grafting Stem Cells from the Adult Human Brain – A Feasible Strategy?

According to the NIH (http://stemcells.nih.gov/info), the adult blood-forming stem cells in bone marrow are currently the only type of stem cell commonly used to treat human diseases, as in bone marrow transplants included in the treatment of leukemia (blood cancer). Transplantations into the brain have also been investigated in clinical trials, but not with stem cells. Since the late 1980s, over 350 patients worldwide have received tissue from the brains of aborted fetuses for Parkinson's disease, with variable symptomatic relief in grafted patients (Freed et al. 2001; Lindvall and Bjorklund 2004). There are, however, a number of problems associated with the use of fetal tissue, including poor survival of transplanted cells, risk of immunological rejection, risk of disabling side effects, difficulties with supply of tissue (material from 3–5 embryos needed in each side of the brain for therapeutic effect) and ethical concerns regarding the use of tissue from aborted human fetuses (Galvin and Jones 2006). In a similar manner, the use of human embryonic stem cells is also beset by ethical concerns, as well as concerns regarding potential tumor formation (Odorico et al. 2001; Bjorklund et al. 2002).

Autologous transplantation (i.e. cells from one patient grafted back into the same patient) could circumvent a number of these problems, using stem cells from a patient's own brain (Taupin and Gage 2002; Langmoen et al. 2003; Galvin and Jones 2006; Taupin 2006). It has been shown that such cells can be harvested from the brain by surgery, either via endoscopic techniques or as tissue fragments harvested during surgery for epilepsy, and furthermore that cells can be grown from these sources in quantities theoretically sufficient to treat patients with Parkinson's disease (Westerlund et al. 2003; Moe et al. 2005a, b; Westerlund et al. 2005).

Little is known, however, about the properties of grafted adult neural stem cells, especially regarding such cells from the human brain. Very few studies have investigated the properties and the behavior of adult human neural stem cells following transplantation into the diseased or injured brain (Galvin and Jones 2006). Steindler and colleagues demonstrated unprecedented plasticity in cells derived from the adult human brain, and showed robust cell survival following transplantation into intact mouse brains (Walton et al. 2006). However, they did not include the important element of some form of brain injury and the influence this might have on survival, migration and differentiation of the grafted cells. Another study showed eight weeks survival of oligodendrocyte-precursors from adult human brain grafted into adult rat brains, accompanied by high doses of immunosuppressant drugs (Windrem et al. 2002).

Following our studies demonstrating that a single, neural stem cell from the adult brain can develop functional neuronal networks in a culture dish (Moe et al. 2005a, b), we wanted to find out whether these stem cells could be transplanted into an adult brain with an ischemic lesion, respond to the injury and develop into more mature brain cells. We showed that the transplanted cells had a striking ability to migrate towards and into the injury and started to mature in glial and neuronal directions, without any sign of tumor formation. However, many of the transplanted cells remained immature, and also a significant number of cells died. This was probably to a certain degree a result of the immune system of the rat trying to break down the foreign human cells, despite the fact that the rats received medicine to suppress this effect. The brain used to be viewed as an 'immunologically privileged site', meaning that the immune system will not attack transplanted cells, but this has proven to be wrong (Barker and Widner 2004). One could speculate that cell survival would be improved with an autologous graft, as indicated in a previous study in rats (Duan et al. 1995). On the other hand, Toda and colleagues reported relatively poor survival after transplantation of cells within an inbred strain of rats (Toda et al. 2001), indicating that other factors are of importance in graft survival. These factors apply to cell transplantations in general, and include cell injury during the procedure, insufficient nutrition, oxidative stress and lack of trophic support in the host (Le Belle et al. 2004).

We are currently investigating the possibility of manipulating the stem cells before the transplantation, in order to acquire cells with a greater ability to mature in the host brain. Other groups have done such a 'pre-differentiation' before grafting of cells of embryonic origin, resulting in the generation of more neurons in the host (Wu et al. 2002). In general, human adult neural stem cells are more difficult to culture, and therefore it is also more challenging to control their fate and create specific types of neurons, for example dopaminergic cells (Galvin and Jones 2006).

4.5 Conclusion

The old dogma of no neurogenesis in the adult human brain is no longer valid. Indeed, new neurons are generated in certain parts of the brain, and can also be induced in other parts of the brain as a response to injury. These facts raise the possibility of novel therapies for brain injuries and neurodegenerative diseases, either by stimulating and recruiting the endogenous stem cells or by transplantation of these cells into the adult brain. As such, it represents an alternative to the use of fetal and embryonic material, which is beset by ethical concerns and higher risk of tumor formation.

Neural stem cells can be harvested from the adult human brain, multiply in culture and develop into functional neuronal networks. Autologous cell replacement is therefore an exciting possibility, where neural stem cells from a patient are expanded in culture and transplanted into the same patient's brain. However, there are currently a number of limitations to this approach, including concerns about the possibility for long-term culturing, the generation of specific types of neurons from these cells and the behavior following transplantation in animal models of diseases. Furthermore, cells from a patient may harbor genetic flaws or predisposition to the disease in question, or they may be functionally impaired due to age or long-term drug treatment (Snyder and Olanow 2005). The limitations are to a large degree based on a lack of knowledge of both adult human neural stem cell biology and of the exact mechanisms underlying diseases in the brain, making it difficult at present to predict their usefulness in future cell replacement therapies. Comprehensive preclinical testing is still required before the full potential of adult human neural stem cells can be realized.

References

Altman, J. (1969). "Autoradiographic and histological studies of postnatal neurogenesis. IV. Cell proliferation and migration in the anterior forebrain, with special reference to persisting neurogenesis in the olfactory bulb", *J Comp Neurol* 137(4): 433–57.

Altman, J. and G. D. Das (1965). "Autoradiographic and histological evidence of postnatal hippocampal neurogenesis in rats", *J Comp Neurol* 124(3): 319–35.

Arsenijevic, Y., J. G. Villemure, et al. (2001). "Isolation of multipotent neural precursors residing in the cortex of the adult human brain", *Exp Neurol* 170(1): 48–62.

Arvidsson, A., T. Collin, et al. (2002). "Neuronal replacement from endogenous precursors in the adult brain after stroke", *Nat Med* 8(9): 963–70.

Backlund, E. O. (1987). "Transplantation to the brain – a new therapeutic principle or useless venture?", *Acta Neurochir Suppl* (Wien) 41: 46–50.

Backlund, E. O., P. O. Granberg et al. (1985). "Transplantation of adrenal medullary tissue to striatum in parkinsonism. First clinical trials", *J Neurosurg* 62(2): 169–73.

Barker, R. A. and H. Widner (2004). "Immune problems in central nervous system cell therapy", *NeuroRx* 1(4): 472–81.

Bjorklund, L. M., R. Sanchez-Pernaute, et al. (2002). "Embryonic stem cells develop into functional dopaminergic neurons after transplantation in a Parkinson rat model", *Proc Natl Acad Sci USA* 99(4): 2344–9.

Brockes, J. P. (1997). "Amphibian limb regeneration: rebuilding a complex structure", *Science* 276(5309): 81–7.

Duan, W. M., H. Widner, et al. (1995). "Temporal pattern of host responses against intrastriatal grafts of syngeneic, allogeneic or xenogeneic embryonic neuronal tissue in rats", *Exp Brain Res* 104(2): 227–42.

Emsley, J. G., B. D. Mitchell, et al. (2005). "Adult neurogenesis and repair of the adult CNS with neural progenitors, precursors, and stem cells", *Prog Neurobiol* 75(5): 321–41.

Eriksson, P. S., E. Perfilieva, et al. (1998). "Neurogenesis in the adult human hippocampus", *Nat Med* 4(11): 1313–7.

Freed, C. R., P. E. Greene, et al. (2001). "Transplantation of embryonic dopamine neurons for severe Parkinson's disease", *N Engl J Med* 344(10): 710–9.

Gage, F. H. (2000). "Mammalian neural stem cells", *Science* 287(5457): 1433–8.

Galvin, K. A. and D. G. Jones (2006). "Adult human neural stem cells for autologous cell replacement therapies for neurodegenerative disorders", *NeuroRehabilitation* 21(3): 255–65.

Goh, E. L., D. Ma, et al. (2003). "Adult neural stem cells and repair of the adult central nervous system", *J Hematother Stem Cell Res* 12(6): 671–9.

Hablitz, J. J. and I. A. Langmoen (1982). "Excitation of hippocampal pyramidal cells by glutamate in the guinea- pig and rat", *J Physiol* 325: 317–31.

Johansson, C. B., S. Momma, et al. (1999). "Identification of a neural stem cell in the adult mammalian central nervous system", *Cell* 96(1): 25–34.

Johansson, C. B., M. Svensson, et al. (1999). "Neural stem cells in the adult human brain", *Exp Cell Res* 253(2): 733–6.

Kelly, S., T. M. Bliss, et al. (2004). "Transplanted human fetal neural stem cells survive, migrate, and differentiate in ischemic rat cerebral cortex", *Proc Natl Acad Sci USA* 101(32): 11839–44.

Kempermann, G., L. Wiskott, et al. (2004). "Functional significance of adult neurogenesis", *Curr Opin Neurobiol* 14(2): 186–91.

Kornblum, H. I. (2007). "Introduction to neural stem cells", *Stroke* 38(2 Suppl): 810–6.

Kukekov, V. G., E. D. Laywell, et al. (1999). "Multipotent stem/progenitor cells with similar properties arise from two neurogenic regions of adult human brain", *Exp Neurol* 156(2): 333–44.

Langmoen, I. A., M. Ohlsson, et al. (2003). "A new tool in restorative neurosurgery: creating niches for neuronal stem cells", *Neurosurgery* 52(5): 1150–53.

Le Belle, J. E., M. A. Caldwell, et al. (2004). "Improving the survival of human CNS precursor-derived neurons after transplantation", *J Neurosci Res* 76(2): 174–83.

Lie, D. C., H. Song, et al. (2004). "Neurogenesis in the adult brain: new strategies for central nervous system diseases", *Annu Rev Pharmacol Toxicol* 44: 399–421.

Liker, M. A., G. M. Petzinger, et al. (2003). "Human neural stem cell transplantation in the MPTP-lesioned mouse", *Brain Res* 971(2): 168–77.

Lindvall, O. and A. Bjorklund (2004). "Cell therapy in Parkinson's disease", *NeuroRx* 1(4): 382–93.

Lopez-Garcia, C., A. Molowny, et al. (1992). "Lesion and regeneration in the medial cerebral cortex of lizards", *Histol Histopathol* 7(4): 725–46.

Martino, G. and S. Pluchino (2006). "The therapeutic potential of neural stem cells", *Nat Rev Neurosci* 7(5): 395–406.

Mezey, E., K. J. Chandross, et al. (2000). "Turning blood into brain: cells bearing neuronal antigens generated in vivo from bone marrow", *Science* 290(5497): 1779–82.

Moe, M. C., M. Varghese, et al. (2005a). "Multipotent progenitor cells from the adult human brain: neurophysiological differentiation to mature neurons", *Brain* 128(Pt 9): 2189–99.

Moe, M. C., U. Westerlund, et al. (2005b). "Development of neuronal networks from single stem cells harvested from the adult human brain", *Neurosurgery* 56(6): 1182–8; discussion 1188–90.

Nakatomi, H., T. Kuriu, et al. (2002). "Regeneration of hippocampal pyramidal neurons after ischemic brain injury by recruitment of endogenous neural progenitors", *Cell* 110(4): 429–41.

Nottebohm, F. (1981). "A brain for all seasons: cyclical anatomical changes in song control nuclei of the canary brain", *Science* 214(4527): 1368–70.

Nottebohm, F. (1985). "Neuronal replacement in adulthood", *Ann N Y Acad Sci* 457: 143–61.

Nottebohm, F. and A. P. Arnold (1976). "Sexual dimorphism in vocal control areas of the songbird brain", *Science* 194(4261): 211–3.

Nunes, M. C., N. S. Roy, et al. (2003). "Identification and isolation of multipotential neural progenitor cells from the subcortical white matter of the adult human brain", *Nat Med* 9(4): 439–47.

Odorico, J. S., D. S. Kaufman, et al. (2001). "Multilineage differentiation from human embryonic stem cell lines", *Stem Cells* 19(3): 193–204.

Olstorn, H., M. C. Moe, et al. (2007). "Transplantation of stem cells from the adult human brain to the adult rat brain", *Neurosurgery* 60(6): 1089–98; discussion 1098–9.

Ostenfeld, T., M. A. Caldwell, et al. (2000). "Human neural precursor cells express low levels of telomerase in vitro and show diminishing cell proliferation with extensive axonal outgrowth following transplantation", *Exp Neuro* 164(1): 215–26.

Pessina, A. and L. Gribaldo (2006). "The key role of adult stem cells: therapeutic perspectives", *Curr Med Res Opin* 22(11): 2287–300.

Pluchino, S., A. Quattrini, et al. (2003). "Injection of adult neurospheres induces recovery in a chronic model of multiple sclerosis", *Nature* 422(6933): 688–94.

Ramon y Cajal, S. (1913). "Degeneration and regeneration of the nervous system", (London, Oxford UP, 1928.): (Day RM, translator, from the 1913 Spanish edition).

Reynolds, B. A. and S. Weiss (1992). "Generation of neurons and astrocytes from isolated cells of the adult mammalian central nervous system", *Science* 255(5052): 1707–10.

Riess, P., C. Zhang, et al. (2002). "Transplanted neural stem cells survive, differentiate, and improve neurological motor function after experimental traumatic brain injury", *Neurosurgery* 51(4): 1043–52; discussion 1052–4.

Roy, N. S., S. Wang, et al. (2000). "In vitro neurogenesis by progenitor cells isolated from the adult human hippocampus", *Nat Med* 6(3): 271–7.

Snyder, B. J. and C. W. Olanow (2005). "Stem cell treatment for Parkinson's disease: an update for 2005", *Curr Opin Neurol* 18(4): 376–85.

Song, H. J., C. F. Stevens, et al. (2002). "Neural stem cells from adult hippocampus develop essential properties of functional CNS neurons", *Nat Neurosci* 5(5): 438–45.

Svendsen, C. N. and M. A. Caldwell (2000). "Neural stem cells in the developing central nervous system: implications for cell therapy through transplantation", *Prog Brain Res* 127: 13–34.

Svendsen, C. N., M. A. Caldwell, et al. (1997). "Long-term survival of human central nervous system progenitor cells transplanted into a rat model of Parkinson's disease", *Exp Neurol* 148(1): 135–46.

Tai, Y. T. and C. N. Svendsen (2004). "Stem cells as a potential treatment of neurological disorders", *Curr Opin Pharmacol* 4(1): 98–104.

Taupin, P. (2006). "Autologous transplantation in the central nervous system", *Indian J Med Res* 124(6): 613–8.

Taupin, P. and F. H. Gage (2002). "Adult neurogenesis and neural stem cells of the central nervous system in mammals", *J Neurosci Res* 69(6): 745–9.

Thored, P., A. Arvidsson, et al. (2005). "Persistent production of neurons from adult brain stem cells during recovery after stroke," *Stem Cells*.

Toda, H., J. Takahashi, et al. (2001). "Grafting neural stem cells improved the impaired spatial recognition in ischemic rats", *Neurosci Lett* 316(1): 9–12.

van Praag, H., A. F. Schinder, et al. (2002). "Functional neurogenesis in the adult hippocampus", *Nature* 415(6875): 1030–4.

Walton, N. M., B. M. Sutter, et al. (2006). "Derivation and large-scale expansion of multipotent astroglial neural progenitors from adult human brain", *Development* 133(18): 3671–81.

Wennersten, A., X. Meier, et al. (2004). "Proliferation, migration, and differentiation of human neural stem/progenitor cells after transplantation into a rat model of traumatic brain injury", *J Neurosurg* 100(1): 88–96.

Westerlund, U., M. C. Moe, et al. (2003). "Stem cells from the adult human brain develop into functional neurons in culture", *Exp Cell Res* 289(2): 378–83.

Westerlund, U., M. Svensson, et al. (2005). "Endoscopically harvested stem cells: a putative method in future autotransplantation", *Neurosurgery* in press.

Windrem, M. S., N. S. Roy, et al. (2002). "Progenitor cells derived from the adult human subcortical white matter disperse and differentiate as oligodendrocytes within demyelinated lesions of the rat brain", *J Neurosci Res* 69(6): 966–75.

Wu, P., Y. I. Tarasenko, et al. (2002). "Region-specific generation of cholinergic neurons from fetal human neural stem cells grafted in adult rat", *Nat Neurosci* 5(12): 1271–8.

Chapter 5
Can We Use Human Embryonic Stem Cells to Treat Brain and Spinal Cord Injury and Disease?

Joel C. Glover

Abstract The potential use of human embryonic stem cells for the treatment of neurological disease and injury is discussed from the perspectives of two common disease scenarios. Spinal cord injury and diseases such as multiple sclerosis that affect specific cell types in the spinal cord represent a substantial proportion of all neuropathologies and are among the most heavily targeted by efforts to establish stem cell-based replacement therapies. Parkinson's disease selectively destroys a single type of neuron in a restricted region of the brain. For this reason it was the first neurological disease for which cell replacement therapy was attempted in humans and is considered one of the most amenable to treatment using stem cells. Although the replacement of a single cell type or the repair of a restricted lesion would appear to be relatively straightforward, several issues conspire to make stem cell-based replacement therapy in the brain and spinal cord substantially more challenging. These include the inherent complexity of neural circuits, the problems of ensuring the survival of stem cells and their derivatives after implantation and directing their differentiation into the appropriate cell types, and the increased refractoriness of chronic injury to treatment due to changes in the cellular environment. A layman's guide to the composition of brain and spinal cord tissue is provided, and an update of recent advances in basic neuroscience and stem cell research with relevance to these issues is presented.

Keywords Amyotrophic lateral sclerosis, demyelinating diseases, motoneuron diseases, multiple sclerosis, Parkinson's disease

5.1 Introduction

A highly profiled potential arena for the clinical use of stem cells is the treatment of neurological disease. This is because tissue regeneration is particularly limited in the nervous system, and loss of neurons and nerve fibers due to injury or disease

Department of Physiology, Institute of Basic Medical Sciences, University of Oslo PB 1103 Blindern, 0317 Oslo, Norway. e-mail: joel.glover@medisin.uio.no

L. Østnor (ed.), *Stem Cells, Human Embryos and Ethics: Interdisciplinary Perspectives.* 55
© Springer Science + Business Media B.V. 2008

can therefore lead to severe and irreversible loss of function. Using stem cells to overcome the inherently limited regenerative capacity of the nervous system would revolutionize neurology, and could bring tremendous benefit to large patient groups suffering from debilitating disorders such as spinal cord injury, demyelinating diseases, movement disorders, Alzheimer's disease, and stroke. But how promising is the stem cell approach? Can we use stem cells to treat any neurological disease, or are there limitations to the capacity of stem cells to recreate the essential neural connections that are lost in these pathologies? If the latter, can we expect to 'push the envelope' sufficiently both in our understanding of the molecular and cellular deficits involved and in our ability to manipulate stem cells so that rational and economically feasible stem cell-based treatments can eventually be designed for neuropathologies? These are pivotal questions on which the future of neural stem cell research hinges. The aim of this article is to provide non-neuroscientists with a perspective on the challenges facing this field, by focusing on the potential application of human embryonic stem cells to the treatment of two specific diseases: spinal cord injury and Parkinson's disease.

The nervous system is extremely complex. Not just in its gross anatomical structure, with a myriad of fiber tracts and highly branched peripheral nerves investing the body with innervation, but also in the degree of cellular interactions on which brain function depends. Each nerve cell, or neuron, is in isolation a powerful analog computational device, so complex that considerable computer power is required to simulate accurately the ability of just one neuron to generate and integrate chemical and electrical signals. The human brain and spinal cord contain on the order of 100 billion neurons and 10–50 times as many glia cells, and many tens or hundreds of thousands of these can be destroyed by a single small lesion. Moreover, neurons come in a bewildering variety of functional types, with characteristic branching shapes, biochemical signalling molecules, and electrical firing patterns. The complexity becomes unfathomable when one considers that each neuron is embedded in highly interconnected neural circuits whose specific connectivity patterns are decisive for proper function. Each neuron can make synaptic connections with hundreds of other neurons and receive connections from thousands or tens of thousands of other neurons. The total number of synaptic connections in the human brain has been estimated at 100–1,000 trillion.

Given this complexity, it might seem preposterous to imagine that injecting stem cells into an injured or diseased brain could reinstate functions that have been lost. But such pessimism is tempered by the fact that the brain's complexity also engenders it with a remarkable plasticity, a capacity to continuously strengthen and weaken, make and break connections and reorganize circuits so that new functions can be acquired. Increasing knowledge is being gained about how this plasticity is regulated and gives rise to memory and the ability to learn. If stem cells could be used to boost the brain's innate plasticity it could augment and accelerate recovery from neurological damage. Moreover, developmental neuroscientists are gaining increasing insight into how the brain's intricate pattern of connections is established to begin with during embryonic, fetal and early postnatal development. Armed with this information, manipulation of stem cells to generate specific neuron types and form specific synaptic connections is becoming increasingly realistic. Add to this the proposition that in some cases of disease perhaps only a minority of lost connections need to be repaired to recoup function. On this backdrop, and given the imaginative uses of stem cells being pursued, stem cell treatments for damage and disease even of the scope of Alzheimer's and stroke should not be dismissed out of hand.

The use of human embryonic stem cells (hESCs) to treat neurological pathologies is only one of a broad spectrum of potential treatment strategies, all of which are being pursued actively in neuroscience research. It is important to be aware that other approaches may end up being just as powerful as or perhaps even more powerful than using stem cells. A comprehensive account of the issue is beyond the scope of this chapter, but in defining research goals and public policy the use of hESCs must be weighed against at least the following: (1) using exogenous adult (somatic) stem cells, (2) promoting the in situ proliferation and differentiation of endogenous neural stem cells, (3) promoting the inherent capacity for axon regrowth and synaptic plasticity, (4) developing new drugs, (5) developing more effective regimens of training and rehabilitation, (6) developing treatments based on the electrical or chemical stimulation of specific brain structures, such as deep brain stimulation for treating basal ganglia disorders, and (7) brain-machine interfaces and neural prosthetics. One of the great challenges facing modern medicine, in this case neurology, is to ascertain the cost-benefit ratios of the available options for each particular disease.

With respect to stem cells as a source of neurons and glia, both embryonic stem cells and somatic stem cells are being investigated actively world-wide. Embryonic stem cells have the greatest potential, but are subject to both ethical and technical problems. In particular, they pose difficulties with respect to tissue rejection (they are not intrinsically autologous) and tumorigenic potential (which could override the desired direction of differentiation). Somatic stem cells provide a source of autologous, non-tumorigenic cells but may be difficult to generate in sufficient quantities and too restricted in their differentiation potential for all applications. For both sources, highly creative strategies are being pursued to overcome the limitations, such as nuclear reprogramming and altered nuclear transfer.

For a discussion of the different types of stem cells (embryonic, somatic), their potential sources, and strategies for manipulating their properties, see the chapters by Funderud, Borge, Ølstørn et al., and Hurlbut.

5.2 Spinal Cord Injury

Spinal cord injury due to trauma is a leading cause of disability throughout the world, with an incidence of over 10,000 new cases per year in Europe alone. The incidence is unlikely to decline in the near future, as most cases arise in young adults through accidents, in particular traffic accidents. The rising use of automobiles in China and other rapidly developing Asian countries portends a huge number of new cases in that part of the world during the coming decades. The palliative treatments currently available are costly. In the U.S., the lifetime cost for a single patient with a life expectancy of 60 years after injury has been estimated at 4–12 million dollars, depending on the severity. The economics provide a strong motivation for developing curative treatments among which stem cell strategies are prominent.

5.2.1 What is the Spinal Cord Made of?

The spinal cord is an extension of the brain (Fig. 5.1a) and as such contains the same kind of tissue with the same kinds of cells as does the brain. The general complexity is, however, somewhat lower, in the sense that the spinal cord contains neural circuits with relatively automatic functions that are better understood than many of the higher functions in the brain itself. For example, the spinal cord contains circuits that mediate the many reflexes that are important for reacting properly to the environment (for example the withdrawal reflex, in which you pull away from a painful stimulus, or the stretch reflex, which helps coordinate movements) and for regulating the state of the body (reflexes that control internal organs), as

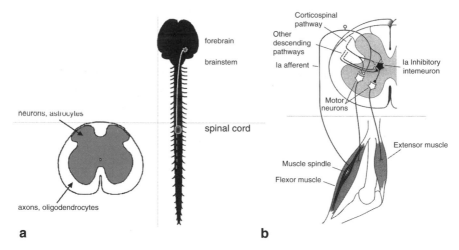

a   b

Fig. 5.1 a. A schematic view of the brain and spinal cord showing the forebrain (location of premotor neurons in the motor cortex, the so-called 'upper motoneurons'), the brain stem (location of many neuron populations with descending axons to the spinal cord), and the spinal cord (in which an injury affecting a neuron with a descending axon is shown). To the left, a transverse section through the spinal cord showing the core of grey matter (where virtually all neurons and astrocytes are found) and the surrounding white matter (where longtiudinal axons and the oligodendrocytes that invest them with myelin are found). **b.** A highly simplified schematic representation of a spinal reflex circuit that controls limb muscle contraction. Shown are the motoneurons that innervate the muscles, a sensory neuron (Ia afferent) that relays information about muscle contraction back into the spinal cord, and an interneuron and multiple descending synaptic inputs from the brain that regulate the activity of the motoneuron. Panel b from Principles of Neuroscience, 3rd edition, 1991, edited by Kandel et al., Springer, with permission.

well as circuits that generate relatively automatic and repetitive movements, such as walking. But even though the functions subserved are relatively easy to understand, many of the circuits involved are still complex enough to defy adequate description today. Much more research needs to be done before a full account of the types of neurons involved in these spinal cord circuits can be obtained. This is one of the main challenges facing stem cell strategies: not knowing precisely what needs to be replaced after an injury.

The neural circuits in the spinal cord are composed of three basic types of neuron (Fig. 5.1b). *Motor neurons*, sometimes called *motoneurons*, are special because they send their nerve fibers, also known as *axons*, out of the spinal cord to innervate muscles or peripheral neurons in the body. *Interneurons* are interconnected with each other and with motoneurons in such a way that they can drive the motoneurons in particular patterns of activity. For example, the interneurons that are responsible for the walking circuit are connected so that muscles in the two legs are activated in alternation, so that the legs make their steps in alternation. This activity pattern is imposed onto the motoneurons which then transmit it to the muscles (Kiehn 2006). Motoneurons and many interneurons also receive sensory input from *sensory neurons*, whose cell bodies lie outside the spinal cord but whose axons extend into the spinal cord. The signals which the sensory neurons transmit into the spinal cord convey information about what is happening in and around the body so that the activity of the interneurons and motoneurons can be modulated appropriately. For example, during walking, one might trip over a small unevenness in the sidewalk. The disruption of the intended movement is picked up by sensors in the legs and in the organs of balance and the information is transmitted by the sensory neurons to the interneurons and motoneurons in the spinal cord so that they change their activity patterns to compensate for the disturbance. This happens automatically (reflexively) so that the brain itself doesn't have to take conscious action (although the event would certainly be registered consciously as having happened).

Interneurons, like motoneurons, have axons, but unlike motoneuron axons these do not exit the spinal cord (although the axons of some interneurons, called *projection neurons*, extend all the way up into the brain). Some interneuron axons are short, others long, some stay on one side of the spinal cord, others cross to the other side, and some extend up the spinal cord while others extend down. The interneurons can be classified according to where their axons go, which other neurons (interneurons and motoneurons) they connect with, what kind of sensory information they listen to, what kind of intrinsic firing patterns and chemical messengers they employ, and even which genes they express (Goulding et al. 2002; Nissen et al. 2005). We do not yet know how many different types of interneurons exist in the spinal cord, but a conservative estimate based on the attributes listed above would be on the order of several tens to a hundred. These are distributed differentially within the spinal cord in an overall anatomical pattern that is still only partially understood (Goulding et al. 2002; Nissen et al. 2005). Nevertheless, the locations of certain circuits have been pinpointed, such that we know for example that the circuit that controls walking is located in the upper lumber segments of the spinal cord, even though we don't know exactly which interneurons make up that circuit (Kiehn 2006).

Motoneurons are functionally a much more homogeneous population of neurons than the interneurons, but they also differ according to whether they innervate muscles or peripheral neurons, and also according to which muscles they innervate. Their distribution is quite well characterized, so we know where the motoneurons that innervate a given muscle are located along the length of the spinal cord.

In addition to motoneurons and interneurons, another class of cells, the *glial cells,* are present in the spinal cord (Fig. 5.2). Glial cells have for many years been considered to provide structural and metabolic support for neurons, but more recent research indicates that some of them also send signals among themselves and in a limited fashion also to neurons, so the spectrum of functional roles these cells possess is being extended. There are three basic types of glial cells, each of which has specific, well-known functions. *Astrocytes* regulate the cellular environment around the nerve cells and contribute to the blood-brain barrier, *oligodendrocytes* form the electrical insulation called *myelin* around nerve fibers, and *microglia* participate in inflammatory reactions.

If one cuts transversely through the spinal cord and looks at the cut end, one sees a fundamental anatomical feature that is highly relevant for injury and disease scenarios, namely that there is a central area of slightly darker tissue that is surrounded by a lighter rind (Fig. 5.1a). The central core is called the grey matter and is where nearly all of the neurons and astrocytes are located, and the lighter, outer rind is called the white matter and is where most of the nerve fibers are located, including all the nerve fibers that carry information from the brain to the spinal cord and vice versa. The whiteness of the white matter is due to the high fat content of the myelin insulation around the nerve fibers. Because of this organization, trauma or disease that is limited to the white matter will affect primarily nerve fibers, whereas any insult to the grey matter will affect the neurons. Some injuries and diseases will of course impact on both.

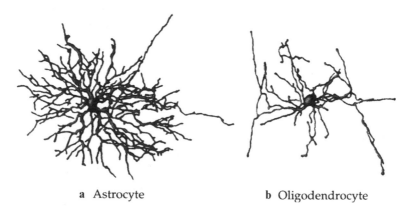

a Astrocyte   b Oligodendrocyte

Fig. 5.2 The astrocytes and oligodendrocytes of the brain and spinal cord, revealed by immuno-histochemical procedures that label the two glial cell types differentially. From 'Neuroscience', 1st edition, 1997, edited by Purves et al., Sinauer, with permission.

There are on the order of 100 to a few thousand motoneurons innervating each muscle in the body, depending among other things on the size of the muscle, and there are on the order of 650 muscles in the body (although not all are innervated by motoneurons in the spinal cord, some are innervated by motoneurons located in the brain stem) so the total number of motoneurons in the human spinal cord is not likely to exceed a million. In fact, in rats and mice the number of spinal motoneurons has been determined fairly accurately to be about half a million and a third of a million, respectively (Bjugn and Gundersen 1993; Bjugn 1993). In rats and mice there are believed to be 10–12 times as many interneurons as motoneurons. Estimates of the total number of neurons in the human spinal cord go as high as 13 million. On top of this there are about twice as many glial cells as neurons. The white matter of the spinal cord contains several million axons, some of which derive from spinal neurons but many of which derive from neurons in the brain.

The adult human spinal cord is a little under a half a meter long and about 1.5 cm in diameter. Given the numbers above, even a small nick in the white matter could sever many tens of thousands of axons and an injury that destroys a half a centimeter of the spinal cord could eliminate 100,000 neurons. It is therefore not surprising that spinal cord injuries can have such catastrophic effects.

5.2.2   What Happens?

What actually happens after a traumatic injury to the spinal cord? As in a wound to any tissue, the initial trauma will injure or kill cells at the trauma site. Thus, a certain number of neurons and glial cells will be eliminated, and a certain number of axons will be severed (but their parent neurons, which may be located quite some distance away, will very likely survive). But the trouble has really only just begun. Over the course of hours to a few days, secondary damage will spread from the primary injury site to affect surrounding tissue (Fig. 5.3). The secondary damage, which is most pronounced during the first 24 hours after injury, is due to several factors, including loss of blood supply and oxygen due to damaged blood vessels, pressure damage due

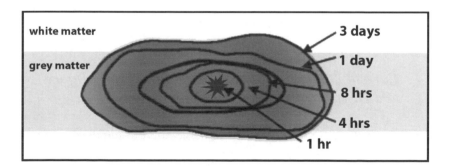

Fig. 5.3 The time course of the spread of secondary damage following a traumatic spinal cord injury

to intraspinal bleeding, the release of ions, neurotransmitters and chemical factors from damaged cells that can provoke injury and death in other cells in the vicinity, and the initial stages of inflammatory responses. Thus, an initial injury that destroys a small fraction of a cubic centimeter of the spinal cord can spread to destroy many cubic centimeters of tissue within a few days. For this reason, one of the main strategies being pursued for the acute treatment of spinal cord injury is to try to limit the secondary damage, for example by cooling, which has been used successfully to limit secondary damage in the brain after stroke and ischemic insults.

As the spinal cord injury progresses through the acute and subacute phases, further changes occur as the inflammatory response triggers astrocytes to enter a reactive state in which they proliferate and generate scar tissue, much as fibroblasts do in other tissues of the body. The astrocytic scar can be quite large and is one of the major impediments to the regeneration of axons in the injured spinal cord, since the scar tissue not only creates a mechanical barrier but is also directly inhibitory to axon growth.

Because of the spread of tissue damage due to secondary injury and the creation of astrocytic scar tissue, the outlook for treating spinal cord injury irrespective of treatment strategy is obviously best for acute injuries and worst for chronic injury cases. In animal experiments, it is commonly seen that success is greater when treatments are initiated soon after an injury (Thuret et al. 2006). Any attempts at using stem cells to treat spinal cord injuries in chronic patients will therefore have to deal simultaneously with the special problems that the chronic situation poses.

5.2.3   Other Types of Spinal Injury and Disease Relevant for Stem Cell-Based Treatment Strategies

A number of other pathological scenarios lead to spinal cord injury and disease in which neurons and glial cells are destroyed. Ischemic insults due to an interruption of blood supply and pressure exerted by tumor metastases can both destroy spinal cord tissue indiscriminately in the absence of trauma. Several well known diseases attack specific types of spinal cord cells. For example, demyelinating diseases such as multiple sclerosis (MS) destroy the oligodendrocytes that create the insulating myelin around axons, and motoneuron diseases such as spinal muscular atrophy (SMA) and amyotrophic lateral sclerosis (ALS) destroy motoneurons.

5.2.4   Current Status of Efforts to Treat Spinal Cord Injury and Disease with Embryonic Stem Cells

Given the overall complexity of spinal cord tissue, it is not surprising that current efforts at using embryonic stem cells to treat spinal cord injury or disease are focused primarily on pathologies that affect only one type of cell. Chief among these are MS, which targets the oligodendrocyte, and diseases such as ALS, which target the motoneuron. The hope is that if a stem cell-based treatment strategy can be developed for these relatively simple cell replacement scenarios, then this may pave the way for developing strategies for more complex scenarios, such as traumatic injuries that can affect all the cell types in the spinal cord.

5.2.4.1 Oligodendrocyte Replacement in Demyelinating Diseases

Among the several demyelinating diseases that have been described, MS is one of the best known among the general public, and one of the most prevalent diseases affecting the central nervous system, with an estimated 2.5 million MS patients world-wide. MS can strike at a variety of ages, as early as late childhood, but typically in young adults (mean age of onset is around 30 years). Women are affected about 50% more often than men. Although a chronic disease, MS need not exhibit incessant progression, but often goes through cycles of remission. It does not affect life span significantly. The basic histopathological scenario is a focal loss of the myelin sheaths that invest nearly all axons in the spinal cord, creating a distinctive plaque that can be observed in the living patient with non-invasive imaging techniques. The number and distribution of plaques determine the loss of function in any given patient. There is no known cure and treatments are purely palliative. Since MS appears to involve an autoimmune response, immune-suppression therapy is becoming an important strategy for controlling symptoms.

Clearly, since the myelin loss associated with MS is focal, replacement strategies can be envisioned in which oligodendrocyte-producing stem or progenitor cells are injected into a plaque, proliferate, and re-invest the area with myelin-producing oligodendrocytes. For those patients with a limited number of isolated plaques this kind of approach could alleviate the majority of symptoms. Since spinal cord injury due to trauma also leads to demyelination, such a treatment would also be beneficial for spinal cord injury patients. How close are we to establishing such a treatment?

During the course of embryogenesis, embryonic stem cells and their descendent lineages are instructed by precisely timed and localized molecular signals which direct differentiation into the proper cell types in the proper locations. In using embryonic stem cells to create a specific cell type such as the oligodendrocyte, it is therefore necessary to know which signals are involved and how these regulate the expression of the genes that characterize that cell type. Great progress has been made recently in understanding the molecular control of the genetic program of oligodendrocyte differentiation (Kitada and Rowitch 2006; Liu et al. 2007). Using this information, researchers have successfully treated hESCs in vitro with molecular factors that promote the generation of oligodendrocytes. Some studies report oligodendrocyte differentiation to over 90% purity, although this rate of success is controversial (Keirstead et al. 2005; Duncan 2005). Human ESC-derived oligodendrocyte progenitors injected into the injured spinal cords of adult rats survive, differentiate into oligodendrocytes, enhance remyelination of denuded axons and promote some recovery of motor function (Keirstead et al. 2005). Unfortunately, this has only been seen in acutely, not chronically injured spinal cords. These findings clearly demonstrate the potential for using hESCs to replenish functional oligodendrocytes. Nevertheless, a number of challenges remain. These include the problem of scaling up from rats to humans (the volume of spinal cord lesions is substantially smaller in rats), the problem of allogenicity (hESC-derived cells could trigger immune reactions and tissue rejection), overcoming the refractoriness of chronic injuries to this and other treatments, and in particular the issue of potential heterogeneity in hESC-derived cells. Since ESCs can in principle generate any cell type, anything less than 100% purity of the desired cell type could introduce undesirable and potentially deleterious side effects. For example, ESCs can potentially generate tumor cells, which would absolutely contraindicate their use in cell replacement therapy.

Despite the obvious challenges, efforts are now being made to generate hESC-derived oligodendrocyte-producing progenitor cells in large numbers for human clinical trials, and within a few years it should become clear whether the promise demonstrated in animal experiments will bear fruit in a clinical setting.

5.2.4.2 Motoneuron Replacement in Motoneuron Diseases

Several diseases target motoneurons and lead to their destruction. SMA presents congenitally or in early childhood and has an incidence of about 1 per 5,000. Some forms of SMA are lethal within a few years whereas others permit survival to adulthood. Adult onset motoneuron diseases include ALS, which has an incidence of 1–2 per 100,000, and which affects not only motoneurons in the spinal cord and brain stem but also premotor neurons (the so-called upper motoneurons) in the motor cortex. ALS has high mortality, with few patients surviving more than 5 years.

Because motoneurons are a relatively homogeneous neuron population, they provide a particularly tractable target for stem cell-based therapies. As is the case for oligodendrocytes, great strides have been made recently in understanding the genetic program of motoneuron differentiation (Shirasaki and Pfaff 2002). Researchers have used this information to direct the differentiation of embryonic stem cells from mice into motoneuron progenitors and from these into functional motoneurons in vitro (Wichterle et al. 2002; Miles et al. 2004). When injected into the spinal cords of rats that have been infected with a virus that kills motoneurons, these embryonic stem cell-derived motoneurons can replace the missing motoneurons, and given appropriate manipulations to induce axon outgrowth, can reinnervate muscle and provide recovery of motor function (Deshpande et al. 2006). Although the overall treatment strategy is complex (new motoneurons must be injected, axon growth inhibitors must be counteracted with drugs, and axon attractants must be seeded into the limbs to coax the motoneuron axons to the muscles), these results clearly demonstrate the potential of using embryonic stem cells to produce functional motoneurons for replacement therapy.

As for oligodendrocytes, despite the promise, substantial challenges remain. First, the results obtained so far using mouse ESCs need to be replicated using hESCs. Thereafter, a number of technical problems need to be solved before the same approach can be used in human patients. Motoneuron diseases, unlike the focal damage to oligodendrocytes caused by MS, can destroy motoneurons throughout the brain stem and spinal cord, and even premotor neurons in the cerebral cortex. One challenge will be to design replacement strategies that facilitate the injection of motoneurons and premotor neurons into such an extensive target. The complex treatment strategy necessary to induce the axons of ESC-derived motoneurons to grow from the spinal cord to the muscles needs to be scaled up, as the distances involved are much greater in humans than in rodents. A major challenge will be to direct the establishment of synaptic connections from interneurons onto the new motoneurons in patterns that produce appropriate function. In addition, some of the same general challenges as noted above for oligodendrocyte replacement (allogenicity, purity, tumorigenicity) need to be addressed. Nevertheless, the generation of motoneurons from hESCs appears likely to be realized in the near future and given the devastating outcome of ALS will almost certainly prompt clinical trials.

5.3 Parkinson's Disease

Parkinson's disease is one of a variety of diseases affecting the basal ganglia deep within the brain. The basal ganglia are large collections of neurons that are involved primarily in the learning, selection and regulation of voluntary movements (although they also contribute to emotional and cognitive functions). Diseases of the basal ganglia therefore present as movement disorders, either as too much movement (such as the uncontrolled movements of dystonias and Huntington's chorea) or too little movement (such as the bradykinesia of Parkinson's disease). Parkinson's disease is remarkable in that it selectively attacks a very specific population of neurons that use dopamine as neurotransmitter and which are located in a structure called the substantia nigra. The substantia nigra provides a diffuse dopaminergic innervation to structures in the basal ganglia, and the mere loss of this dopaminergic innervation creates a critical imbalance of basal ganglia activity that results in the classical symptoms of Parkinson's disease: bradykinesia, rigidity, and tremor (Fig. 5.4). Indeed, the principal treatment for Parkinson's disease today is the medicinal replacement of dopamine by ingestion of the dopamine precursor L-dopa (other treatments, such as surgical lesions or stimulation of specific sites within the basal ganglia with implanted electrodes, are also used). Parkinson's disease and forms of parkinsonism (insults that affect the substantia nigra less specifically and thus produce the same symptoms as Parkinson's disease along with other symptoms) have an incidence of about 1 per 1,000 overall, but this rises to 2–3 per 100 in the elderly (over 70 years of age).

There are only about 200,000 dopaminergic nigral neurons on each side of the brain. Their relative paucity and well-defined, restricted location in the substantia nigra made them a prime target for cell replacement therapy starting already in the late 1970s, when Swedish researchers began implanting embryonic substantia nigra tissue first into the

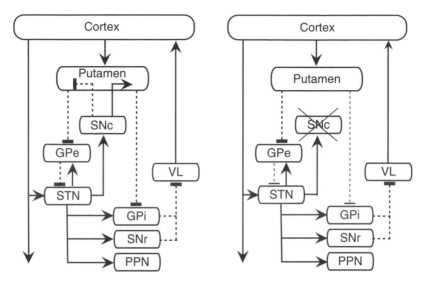

Fig. 5.4 A highly simplified schematic of the connections between different parts of the basal ganglia, the thalamus, and the cortex, along with the dopaminergic inputs to the basal ganglia from the substantia nigra. A. The normally functioning circuit embodies properly balanced excitation (solid lines) and inhibition (dotted lines) so that activity that regulates the selection and initiation of movement by the cortex is appropriate. B. Loss of the dopaminergic inputs to the basal ganglia (through lesion of the substantia nigra pars compacta, SNc) disrupts this balance, resulting in too much inhibition of the thalamus and thus too little activation of movement initiation centers in the cortex. GPe (globus pallidus pars externa), GPi (globus pallidus pars interna), STN (subthalamic nucleus), SNr (substantia nigra pars reticularis), PPN (pedunculopontine nuclei), VL (ventrolateral nucleus of the thalamus). Modified and adapted from Obeso et al. (2000), with permission.

brains of animals with experimentally-induced parkinsonism and then into humans with Parkinson's disease (Björklund and Lindvall 2000). In the animal models, the implanted tissue developed into dopaminergic neurons, some of which innervated the proper basal ganglia targets, and this resulted in the alleviation of the symptoms of the movement disorders exhibited by the animals. Similarly, in human patients treated with substantia nigra tissue from human embryos, signs of dopaminergic innervation have been detected using non-invasive imaging and in postmortem assessment, and clinical improvement above and beyond potential placebo effects have also been documented in some of the patients (Björklund and Lindvall 2000).

The 'proof-of-principle' established by the implantation of embryonic dopaminergic neuron precursors, as well as the ethical issues and practical limitations involved in using tissue directly from human embryos, has prompted intense research into the possibility of using embryonic stem cells to generate dopaminergic neurons. Here there has also been tremendous progress in identifying and characterizing the molecular determinants that establish the dopaminergic phenotype during development. This information has facilitated the recent generation of high-purity dopaminergic neurons with the correct substantia nigra character from mouse ESCs in vitro (Andersson et al. 2006). This advance extends previous work in which hESCs have been induced to differentiate into dopaminergic neurons in vitro, albeit at substantially lower purity, through manipulation of the molecular environment (Taylor and Minger 2005). Implantation of hESC-derived dopaminergic neuron progenitors into animal models of parkinsonism has been shown to alleviate symptoms to some extent, but it has been unclear whether this is the result primarily of a reestablishment of dopaminergic innervation or other effects, such as graft-derived trophic factors or the stimulation of endogenous repair mechanisms. Moreover, survival of hESC-derived dopaminergic neuron progenitors in animal models has been low, indicating that much work remains to be done to ensure that the implanted cells can integrate and survive efficiently. Purity has also been a major issue. Different hESC lines exhibit different capacities for generating dopaminergic neurons, and contaminating cell types can even include non-neural cells that remain proliferative long after the implantation (Taylor and Minger 2005). The hope is that improved protocols, based on greater insight into the normal developmental program for dopaminergic differentiation, can be established for generating high-purity hESC-derived dopaminergic neuron progenitors and that these will provide a better source for implantation.

Thus, although dopaminergic neurons of the substantia nigra provide another attractive target for hESC-based cell replacement therapy, there are still substantial hurdles to cross before a clinically feasible approach is available. In addition to the general problems associated with ESCs, one of the principal difficulties is that the neural circuitry of the basal ganglia and surrounding structures is much less well understood than that of the spinal cord. Moreover, the role of dopamine in regulating basal ganglia function is complex, with different effects being elicited in different basal ganglia target neurons. Loss of dopamine can also trigger compensatory changes in the system (changes in dopamine receptor densities, synaptic rearrangement) that are difficult if not impossible to predict from patient to patient. Recreating a situation in which implanted dopaminergic neurons innervate the basal ganglia in the appropriate way will thus be a major challenge, compounded by the large volume of tissue the basal ganglia occupy in the human brain. Nevertheless, as for oligodendrocytes and motoneurons, it seems highly likely that the generation of pure dopaminergic neuron progenitors from hESCs will be realized. Given advances in our understanding of the way the basal ganglia function, and in how to promote the survival of implanted progenitors, cell replacement therapy will almost certainly become an available option. Whether this option will be competitive in the face of other treatment strategies, which are also advancing rapidly, remains to be seen.

5.4   Alternatives to Cell Replacement

Although the replacement of lost neurons and/or glia is a principal aim of stem cell research, embryonic stem cells may prove to be clinically useful in other, less direct ways. It has been noted in many animal experiments that functional improvement can occur after stem cell implantation despite the lack of obvious neuronal differentation or synaptic integration. Embryonic stem cells and their progeny have been shown to release cytokines and growth factors that are involved in brain development and plasticity (Bentz et al. 2007; Kamei et al. 2007) and in this capacity might promote synaptic plasticity and reorganization without participating directly in the formation of new connections. Treatment with growth factors has been proposed as a potential approach to facilitate recovery from the widespread brain damage associated with stroke and Alzheimer's disease (Kuipers and Bramham 2006). Either as a genetically engineered source of human growth factors in vitro or as an implanted, local in vivo source for targeting growth factor delivery to specific sites in the brain, hESCs could contribute to the development of clinically viable growth factor therapies.

Embryonic stem cells have also been envisioned as a means to provide growth-promoting substrates for damaged axons. A great deal of effort has recently been directed towards using olfactory ensheathing cells for this purpose, including in clinical trials (Thuret et al. 2006). Despite their demonstrated capacity to support axon growth, it is unclear whether olfactory ensheathing cells are the best possible cell type for all areas of the brain and all injury and disease scenarios. Again, because embryonic stem cells can be induced to differentiate into any cell type, they could in principle be engineered to produce growth-promoting cells perfectly tailored to different brain regions and neuronal deficits.

Lastly, embryonic stem cells have been recognized as an ideal source of cells for testing drug therapies in vitro or in animal models prior to embarking on clinical trials. The great diversity of cell types in the brain poses a major challenge for the development of drugs for treating neurological disorders, because drug effects may vary across and within neuronal populations. The possibility of generating unlimited numbers of neurons of any given standardized phenotype for in vitro testing would facilitate a more rapid and reproducible assessment. The same cells could be implanted into animal models to provide testing within an in vivo context.

5.5 Summary

The potential of human embryonic stem cells to generate any cell type within the brain and spinal cord offers hope that cell replacement therapies for a variety of neurological disorders may someday be realized. Stem cell-based replacement strategies are nevertheless highly challenging, due to the complexity of neural tissue and the requirement for a highly controlled differentiation of stem cells. Advances in our understanding of how diverse types of neurons and glial cells differentiate normally are paving the way towards a pre-differentiation of embryonic stem cells into specific lineages prior to use. If the additional challenges of ensuring the survival and functional integration of stem cell-derived neural cells can be met while avoiding side effects, then stem cell-based replacement therapies are likely to become powerful tools in the treatment repertoire. Whether these tools will become economically feasible or competitive in the face of alternative treatment strategies remains to be seen. Embryonic stem cells offer therapeutic promise in other ways as well, for example as potential sources of growth factors or substrates for damaged neurons and axons, and in particular as the source of standardized neural cell populations for drug testing. On this backdrop, continued research into the use of human embryonic stem cells is highly warranted.

References

Andersson, E., Tryggvason, U., Deng, Q., Friling, S., Alekseenko, Z., Robert, B. Perlmann, T., and Ericson, J. (2006). "Identification of intrinsic determinants of midbrain dopamine neurons", *Cell* 124:393–405.

Bentz, K., Molcanyi M., Riess, P., Elbers, A., Pohl, E., Sachinidis, A., Hescheler, J., Neugebauer, E., and Schafer, U. (2007). "Embryonic stem cells produce neurotrophins in response to cerebral tissue extract: cell line-dependent differences", *J Neurosci Res* 85:1057–1064.

Bjugn, R. (1993). "The use of the optical disector to estimate the number of neurons, glial and endothelial cells in the spinal cord of the mouse – with a comparative note on the rat spinal cord", *Brain Res* 627:25–33.

Bjugn, R. and Gundersen, K. (1993). "Estimate of the total number of neurons and glial and endothelial cells in the rat spinal cord by means of the optical dissector", *J Comp Neurol* 328:406–414.

Björklund, A. and Lindvall, O. (2000). "Cell replacement therapies for central nervous system disorders", *Nat Neurosci* 3:537–544.

Deshpande, D. M., Kim, Y. S, Martinez, T., Carmen, J., Dike, S., Shats, I., Rubin, L. L., Drummond, J., Krishnan, C., Hoke, A., Maragakis, N., Shefner, J., Rothstein, J. D., and Kerr, D. A. (2006). "Recovery from paralysis in adult rats using embryonic stem cells", *Ann Neurol* 60:32–44.

Duncan, I. D. (2005). "Oligodendrocytes and stem cell transplantation: their potential in the treatment of leukoencephalopathies", *J Inherit Metab Dis* 28:357–368.

Goulding, M., Lanuza, G., Sapir, T., and Narayan, S. (2002). "The formation of sensorimotor circuits", *Curr Opin Neurobiol* 12:508–515.

Kamei, N., Tanaka, N., Oishi, Y., Hamasaki, T., Nakanishi, K., Sakai, N., and Ochi, M (2007). "BDNF, NT-3, and NGF released from transplanted neural progenitor cells promote corticospinal axon growth in organotypic cocultures", *Spine* 32:1272–1278.

Kiehn, O. (2006). "Locomotor circuits in the mammalian spinal cord", *Annu Rev Neurosci* 29:279–306.

Keirstead, H. S., Nistor, G., Bernal, G., Totoiu, M., Cloutier, F., Sharp, K., and Steward, O. (2005). "Human embryonic stem cell-derived oligodendrocyte progenitor cell transplants remyelinate and restore locomotion after spinal cord injury", *J Neurosci* 25:4694–4705.

Kitada, M. and Rowitch, D. H. (2006). "Transcription factor co-expression patterns indicate heterogeneity of oligodendroglial subpopulations in adult spinal cord", *Glia* 54:35–46.

Kuipers, S. D. and Bramham, C. R. (2006). "Brain-derived neurotrophic factor mechanisms and function in adult synaptic plasticity: new insights and implications for therapy", *Curr Opin Drug Discov Devel* 9:580–586.

Liu, Z., Hu, X., Cai, J., Liu, B., Peng, X., Wegner, M., and Qiu, M. (2007). "Induction of oligodendrocyte differentiation by Olig2 and Sox10: evidence for reciprocal interactions and dosage-dependent mechanisms", *Dev Biol* 302:683–693.

Miles, G. B., Yohn, D. C., Wichterle, H., Jessell, T. M., Rafuse, V. F., and Brownstone R. M. (2004). "Functional properties of motoneurons derived from mouse embryonic stem cells", *J Neurosci* 24:7848–7858.

Nissen, U. V., Mochida, H., and Glover, J. C. (2005). "Development of projection-specific interneurons and projection neurons in the embryonic mouse and rat spinal cord", *J Comp Neurol* 483:30–47.

Obeso, J.A., Rodriguez-Oroz, M.C., Rodriguez, M., Lanciego, J.L., Artieda, J., Gonzalo, N., and Olanow, C.W. (2000) Pathophysiology of the basal ganglia in Parkinson's disease. Trends Neurosci, 23:S8-S19.

Shirasaki, R. and Pfaff, S. L. (2002). "Transcriptional codes and the control of neuronal identity", *Annu Rev Neurosci* 25:251–281.

Taylor, H. and Minger, S. L. (2005). "Regenerative medicine in Parkinson's disease: generation of mesencephalic dopaminergic cells from embryonic stem cells", *Curr Opin Biotechnol* 16:487–492.

Thuret, S., Moon, L. D., and Gage, F. H. (2006). "Therapeutic interventions after spinal cord injury", *Nat Rev Neurosci* 7:628–643.

Wichterle, H., Lieberam, I., Porter, J. A., and Jessell, T. M. (2002). "Directed differentiation of embryonic stem cells into motor neurons", *Cell* 110:385–397.

Chapter 6
Stem Cells, Embryos and Ethics: Is There a Way Forward?*

William B. Hurlbut

Abstract A century of advances in molecular and cell biology have brought us at the dawn of the 21st century back to the study of whole living beings. When these studies are applied to human biology, we must once again consider the fundamental questions about the meaning and significance of nascent human life. The current conflict over embryonic stem cell research is only the beginning of a series of difficult controversies concerning scientific manipulation of human life in its early stages of development. What is needed now is a coherent and reasonable definition of the boundaries of humanity that we seek to defend, one that takes into account the vast wealth of human tradition – social, political and scientific. This would facilitate a resolution that would make possible the advance of science while achieving social consensus. Several proposals have been advanced to achieve this end. One of these methods, altered nuclear transfer, proposes to make use of the technology of somatic nuclear cell transfer, but with a pre-emptive genetic or epigenetic alteration that precludes the integrated and coordinated organization necessary for embryogenesis. The moral and scientific aspects of this proposal are discussed as a way forward for embryonic stem cell research.

Keywords Altered Nuclear Transfer (ANT), embryonic stem cells, pluripotency, totipotency

---

The Neuroscience Institute at Stanford, Stanford University Medical Center, 371 Serra Mall – Stanford University, Room 345, Gilbert Hall, Stanford, CA 94305-5020. e-mail: ethics@stanford. edu

* Portions of this essay are drawn from "Framing the Future: Embryonic Stem Cells, Ethics and the Emerging Era of Developmental Biology", *Pediatrics Research*, Vol 59, No 4, Pt 2, 2006 and "Altered Nuclear Transfer: A Way Forward for Embryonic Stem Cell Research," *Stem Cell Reviews*, Vol 1, 2005.

L. Østnor (ed.), *Stem Cells, Human Embryos and Ethics: Interdisciplinary Perspectives.*
© Springer Science + Business Media B.V. 2008

6.1   Introduction

We are at a crucial moment in the process of scientific discovery. The dramatic advances in molecular biology throughout the 20th century have culminated in the sequencing of the human genome and increasing knowledge of cell physiology and cytology. These studies were accomplished by breaking down organic systems into their component parts. Now, however, as we move on from genomics and proteomics to discoveries in developmental biology, we have returned to the study of living beings. When applied to human biology, this inquiry reopens the most fundamental questions concerning the relationship between the material form and the moral meaning of developing life.

The current conflict over ES cell research is just the first in a series of difficult controversies that will require us to define with clarity and precision the moral boundaries we seek to defend. Human-animal Chimeras, parthenogenesis, projects involving the laboratory production of organs –and a wide range of other emerging technologies will continue to challenge our definitions of human life. These are not questions for science alone, but for the full breadth of human wisdom and experience.

The scientific arguments for going forward with this research are strong.

- The convergence of these advancing technologies is delivering unprecedented powers for research into the most basic questions in early human development.
- Beyond the obvious benefit of understanding the biological factors behind the estimated 150,000 births (in the US alone) with serious congenital defects per year, it is becoming increasingly evident that certain pathologies that are only manifest later in life are influenced or have their origins in early development.
- Furthermore, fundamental developmental processes (including the formation and functioning of stem cells), and their disordered dynamics, seem to be at work in a range of adult pathologies including some forms of cancer.

Yet from the moral and social perspective there are serious concerns. It is important to acknowledge the many scientific projects for which human embryos could be used. Beyond their destruction for the procurement of embryonic cells, some fear the industrial scale production of living human embryos for a wide range of research in natural development, toxicology and drug testing.

Lord Alton, a member of the House of Lords in the UK told me that they estimate over 100,000 human embryos have already been used in scientific experimentation in Britain.

Beyond that, there is concern about the commodification and commercialization of eggs and embryos, and worry about the implications of ongoing research to created an artificial endometrium (a kind of artificial womb) that would allow the extracorporeal gestation of cloned embryos to later stages for the production of more advanced cells, tissues and organs.

Furthermore, from a social and political perspective, the emerging patchwork of policies on the national and international level threatens to create a situation in which there will be commercial motivations for 'outsourcing' ethically controversial or illegal research. Likewise, there will be 'medical tourism' by patients in a desperate quest for cures. And, in countries such as Australia where ESCR is legal but still a matter of controversy, a large percentage of patients may one day enter the hospital with moral qualms about the foundations on which their treatments have been developed. What was traditionally the sanctuary of compassionate care at the most vulnerable and sensitive moments of human life is becoming an arena of controversy and conflict.

Clearly, both sides of this difficult debate are defending important human goods – and both of these goods are important for all of us. A purely political solution will leave our civilization bitterly divided, eroding the social support and sense of noble purpose that is essential for the public funding of biomedical science. Yet, this support is crucial if stem cell science is to advance. In the United States, where there are currently no federally legislated constraints on the use of private funds for this research, there is a consensus opinion in the scientific community that without National Institutes of Health support for newly created embryonic stem cell lines, progress in this important realm of research will be severely constrained.

The current conflict in the political arena is damaging to science, to religion and to our larger sense of cultural unity. The way this debate is proceeding (at least in the United States) is, in my opinion, completely contrary to the positive pluralism that is the strength of our democracy.

In May 2005, acknowledging our nation's impasse over embryonic stem cell research, the President's Council on Bioethics published a 'White Paper' that outlines a series of proposals for obtaining pluripotent stem cells (the functional equivalent of embryonic stem cells) without the creation or destruction of human embryos.[1] In the 18 months since the publication of that report, encouraging scientific progress on each of these approaches has been reported in major scientific journals.

I want to begin by describing these proposals and making some general comments about their problems and prospects. Then I will step back to discuss in some detail the basis for moral objections to embryo-destructive research. Finally, I will draw on a deeper discussion of one of the proposals, Altered Nuclear Transfer, to explore how we might set a moral frame that opens even beyond ESCR into a broader arena of research in developmental biology.

6.2 White Paper

Throughout the Council report a distinction is drawn between *totipotency*, the capacity to give rise to the whole organism as an integrated living being, and *pluripotency*, the capacity to give rise to the many different individual cell types of the human body. A naturally fertilized egg, the one-cell embryo, is totipotent. Embryonic stem cells are merely pluripotent; they lack the immanent powers for self-organization and self-development that characterize an embryonic organism. Each of the four proposals derives its moral justification from this empirically evident distinction.

---

[1] President's Council of Bioethics, *White Paper: Alternative Sources of Human Pluripotent Stem Cells.* (Washington, DC: President's Council of Bioethics, 2005). http:/www.bioethics.gov/reports/white_papier/index.html

6.2.1   Culture of Pluripotent Cells from Embryos
that are Considered Organismically Dead

(This proposal has been put forward by Columbia University physicians Howard Zucker and Donald Landry).

As with the procurement of organs following brain death, this proposal draws on the distinction between the organic parts and the living whole. Individual blastomeres (cells from the pre-blastocyst stage embryo), when removed from failed embryos, may retain the capacity for limited growth and, possibly, for the production of ES cell lines.

The conceptual challenge here is to establish the criteria of 'embryo death,' a cessation of the integrated unity and totipotent capacity that characterize and define a living human embryo. The scientific challenge is to identify a physical marker of this state.

Zucker and Landry have suggested that irreversible arrest of coordinated cell division is an adequate indicator of embryo death. In the adult brain-dead patient, individual cells and even physiological subsystems may continue their biological processes, but these are mere organic momentum; the overarching coherence and coordinated function that characterize a living organism have ceased. Likewise, for an embryo whose defining nature is its capacity for whole development, the irreversible arrest of coordinated cell division indicates an intrinsic deficiency of integrated existence as a living being. For this reason, the pluripotent parts may be removed without violation of moral principle.

Some progress has been reported in this project. Stem cell biologist Miodrag Stojkovic claims success in creating human ESC lines from two blastocyst-stage embryos he considered to be in a state of arrested development. This author, however, failed to describe his criteria for considering the embryos dead.

Meanwhile, Zucker and Landry report some progress in identifying morphological markers that seem to indicate evidence of irreversible arrest in development. Once the validity of these markers in confirmed, further research might lead to some success with this technique.

6.2.2   Culture of Pluripotent Cells from Single Blastomere
Extraction from Living Embryos (Embryo Biopsy)

As with the previous proposal, this project distinguishes the parts from the living whole. Using established techniques of blastomere extraction developed for pre-implantation genetic diagnosis, an individual cell is removed and cultured to form an ES cell line. The goal is to extract this pluripotent part (the individual cell) without damage to the totipotent embryonic organism.

In a controversial paper published on-line in *Nature* in November 2006, stem cell biologist Robert Lanza of Advanced Cell Technologies, a US biotechnology company, claimed success using this method. However, he came under heavy criticism when it was revealed that he actually completely disaggregated the embryos and cultured most of their individual blastomeres – critics objected that he had not actually proved it is possible to extract a single cell and culture it while the rest of the embryo goes on to successful development and birth.

Lanza replied that, technically, his critics were correct, but that it has already been established that embryos can still develop after single cell extraction – so, he said he 'could have done it.' In fact, Lanza was unable to coax a single cell to form a stem cell line unless he co-cultured multiple cells from the same embryo in contiguous wells within the petri dish. It is likely that adjacent cells extracted from the same embryo exchanged essential signals that sustained their development – suggesting that, at the very least, this method needs further refinement.

More fundamentally, however, this approach raises troubling moral questions. Blastomere extraction is usually performed at the eight-cell stage. It is well known from animal studies (in sheep and rabbits) that early cleavage stage blastomeres (including those from the eight-cell stage in some species) may, when separated and cultured, retain (or regain) the capacity to form a full organism. Individual cells from an eight cell human embryo have not been observed to retain totipotency, but this remains a matter of controversy.

Of greater concern, however, is the fact that blastomere extraction may result in the death of the embryo. Although there is disagreement over the actual rate of embryo destruction, the embryo 'endangerment' involved in such a non-therapeutic intervention constitutes a violation of the Dickey Amendment, the US law governing embryo research. Therefore, as currently proposed, neither the embryo biopsy procedure nor research with the cell lines subsequently derived would qualify for federal funding.

This proposal also raises another interesting moral issue. Evidence in animal studies suggests that even at the earliest cleavage stages there are asymmetrical cell divisions and distinct cell fates. Extraction of a single blastomere may not preclude the formation of a whole living organism, but such disruption of early development might lead to a somewhat different or even defective individual. A damaging effect from this procedure, even a severe but subtle one, might not be clearly evident for years or maybe decades – if ever.

6.2.3 Culture of Pluripotent Cells Through the Direct Production of Specifically Engineered Constructs Lacking the Character of Living Embryos – Altered Nuclear Transfer

This proposal, which I will describe in greater detail below, seeks a way to draw on the organic powers of natural developmental dynamics, but without the creation and destruction of living embryos. Whereas the Zucker-Landry plan would extract pluripotent cells from embryos that are no longer totipotent, Altered Nuclear Transfer (ANT) would create 'biological artifacts,' that never rise to the level of totipotency.

This proposal shifts the ethical debate from the question of *when* during development a normal embryo is a human being with moral worth, to the more fundamental question of *what* component parts and organized structure constitute the minimal criteria for considering an entity to be a living human organism.

ANT is a broad concept with a range of possible approaches. It uses the techniques of nuclear transfer, but with a preemptive genetic or epigenetic alteration, to produce a cellular system that lacks the integrated unity and potency of a living being but contains a partial developmental potential capable of generating pluripotent cells.

For ANT the conceptual and moral challenge is the difficult task of defining the boundary between mere cellular growth lacking integrated form and a living human organism. The scientific challenge of ANT is to find the right genetic or epigenetic alteration to ensure that pluripotent cells can be produced while not creating an embryonic human being.

As I will discuss below, proof-of-principle for this approach has been established in studies with mice.

6.2.4   Production of Pluripotent Cells by the Direct Reprogramming of Somatic Cells (Dedifferentiation)

Through a variety of techniques, including fusion of somatic cells with existing ES cells or direct application of ES or oocyte-derived cytoplasmic factors, this proposal seeks to reprogram differentiated cells back to a pluripotent state and thereby bypass entirely the embryonic phase of ES cell development. The scientific premise of this project is that the epigenetic state that characterizes ES cells can be produced even apart from the sequential process of natural embryogenesis by contact with a soup of essential cytoplasmic factors.

Early efforts have yielded some encouraging progress, but there are still technical difficulties to overcome. In a remarkable *tour de force*, Japanese scientist Shinya Yamanaka used viral vectors to inject four genes crucial for pluripotency into differentiated adult body cells. He then boosted production of the protein coded by these genes and the cells took on some the characteristics of embryonic stem cells.

Such a project, which would stop short of creating a totipotent being, seems an ideal solution if it can be successfully accomplished. Nonetheless, even this proposal raises ethical concerns about how one would establish precisely the biochemical boundary between totipotency and pluripotency without the experimental creation and destruction of human embryos.

6.3 Practical Implications of the Proposals

Clearly, each of these proposals offers different practical prospects and comes with distinct moral and scientific challenges.

Proposals #1 (Zucker-Landry) and #2 (Embryo biopsy) carry the convenience of drawing on existing clinical practice of IVF; but this means that the ES cell lines produced by these methods will be limited to the genotypes of available embryos produced with the random genetic recombinations of natural fertilization. This points to a disadvantage of deriving ES cells from IVF embryos that is rarely considered: the genomes of these embryos have never produced fully formed human organisms. New mutations and unpropitious reassortments of genes that are naturally filtered out in the early stages of embryogenesis (like missions scrubbed on the 'launch pad of life') may be undetected in these cell lines. Some lines may be incapable of fully natural cell functions, and could be potentially ineffective or even dangerous as medical agents in cell replacement therapies.

Proposal #3 (Altered Nuclear Transfer) and proposal #4 (dedifferentiation) involve complicated laboratory procedures, but would offer a full range of genotypes using nuclear material with proven organism-forming potential (the donor nucleus). Difficult questions remain concerning epigenetic problems associated with these techniques – these epigenetic problems are evident in the imperfect formation of cloned animals such as Dolly. However, in mouse studies comparing ES cell lines derived from IVF embryos and nuclear transfer, MIT stem cell biologist Rudolf Jaenisch found no evident differences in basic potential (Jaenisch et al. 2006). Preliminary studies with pluripotent cell lines derived by dedifferentiation suggest they too may be fully functional (Eggan et al. 2005).

In either case, using cells of specific genotypes would allow a wider range of research applications (including modeling of disease in chimeric animals, studies of toxicology, and development of genetically tailored pharmaceutical agents) as well as a potential source of immune-compatible cells for transplantation therapies.

Dedifferentiation may prove to be the most technically difficult, but would appear to be the least morally controversial of the four proposals. The basis for its moral acceptability, however, implies a limitation of this approach: it must aim directly at the production of cells with the functional equivalence of ES cells from the blastocyst stage and stay safely clear of producing cells with the totipotent capacities of earlier, cleavage stage blastomeres. This means that scientific study of the molecular processes and cell dynamics of the first five days of development are beyond the reach of such an approach.

The implications of this limitation are evident if we consider the broader significance of the ethical controversy over ESCR. Our current conflict over ESCR is more than an immediate practical problem. It is a symbolic struggle over the whole future of developmental biology – over how we will proceed with a wide range of research on human development. If we are to open this inquiry in a way that will yield the fullest scientific knowledge and medical advance, we must define with clarity and precision the boundaries of our moral principles for the protection of nascent human life.

6.4   Moral Meaning of Emerging Life

Any evaluation of the moral significance of human life must take into account the full procession of continuity and change that is essential for its development. With the act of conception, a new life is initiated with a distinct genetic endowment that organizes and guides the growth of a unique and unrepeatable human being.

The gametes (the sperm and egg), although alive as cells, are not living beings: they are instrumental organic agents of the parents. The joining of the gametes brings into existence an entirely different kind of entity, a living human organism. With regard to fundamental biological meaning (and moral significance), the act of fertilization is a leap from zero to everything.

In both structure and function, the zygote (the single cell embryo) and subsequent embryonic stages differ from all other cells or tissues of the body; they contain within themselves the organizing principle for the full development of a human being. The very word organism implies organization, an overarching principle that binds the parts and processes of life into a harmonious whole. As a living being, an organism is an integrated, self-developing and self-maintaining unity under the governance of an immanent plan.

For an embryonic organism, this implies an inherent potency, an engaged and effective potential with a drive in the direction of the mature form. By its very nature, an embryo is a developing being. Its wholeness is defined by both its manifest expression and its latent potential; it is the phase of human life in which the 'whole' (as the unified organismal principle of growth) precedes and produces its organic parts. The philosopher Robert Joyce explains: 'Living beings come into existence all at once and then gradually unfold to themselves and to the world what they already but only incipiently are (Joyce 1978).' To be a human organism is to be a whole living member of the species *Homo sapiens*, with a human present and a human future evident in the intrinsic potential for the manifestation of the species typical form. Joyce continues: 'No living being can become anything other than what it already essentially is.'

It is this implicit whole, with its inherent potency, that endows the embryo with continuity of human identity from the moment of conception and therefore, from this perspective, inviolable moral status. To interfere in its development is to transgress upon a life in process. The principle of this analysis applies to any entity that has the same potency as a human embryo produced by natural fertilization, regardless of whether it is the product of IVF, cloning or other processes.

This conclusion is consistent with 2,500 years of medical science – as recently as 1948, the Physicians Oath in the Declaration of Geneva, echoing the enduring traditions of Hippocratic medicine, proclaimed: 'I will maintain the utmost respect for human life from the time of conception.'

As we descend into an instrumental use of human life we destroy the very reason for which we were undertaking our new therapies; we degrade the humanity we were trying to heal.

6.5   Altered Nuclear Transfer

If we are to sustain our principles for the protection of human life and at the same time open the broadest possible exploration of the biology processes of embryonic development, we will need to precisely define the boundary between the totipotent living being and the partial organic powers of mere pluripotent cells.

Totipotent: capacity to give rise to the whole organism as an integrated living being.

Pluripotent: capacity to produce all the cell types of the human body but not the coherent and integrated unity of a living organism.

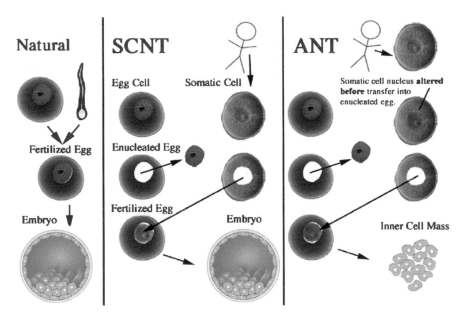

Fig. 6.1   Altered Nuclear Transfer (ANT) differs from both natural fertilization and somatic cell nuclear transfer (SCNT). Natural fertilization combines a sperm and an egg (each with only half of the necessary chromosomes) to generate an embryo with a complete complement of genetic material. In SCNT the nucleus of an egg is removed and replaced by the nucleus of an adult body cell (with the full complement of genetic material). The egg cytoplasm reprograms the adult cell nucleus and embryonic development begins. In ANT, targeted alterations are made in the nucleus of the somatic cell and/or the cytoplasm of the egg before the act of nuclear transfer. These alterations preclude the integrated organization and potential for development that are the defining characteristics of an embryo, and thereby assure that no embryo is created. ANT produces the biological (and moral) equivalent of a tissue culture

Drawing this distinction is central to the proposal for Altered Nuclear Transfer. As outlined in the Council report, Altered Nuclear Transfer (ANT) creates laboratory constructs that never rise to the level of totipotency. Drawing on a subset of the organic powers of natural development, ANT produces 'biological artifacts' capable of forming pluripotent stem cells without the creation and destruction of human embryos.

ANT is a broad concept with a range of possible approaches. It employs the technique of nuclear transfer, but with a preemptive genetic or epigenetic alteration to produce a cellular system that lacks the integrated unity of a living being.

In standard SCNT the nucleus of a differentiated body cell is transferred into an egg cell that has had its own nucleus removed.[2] The egg cytoplasm then reprograms the transferred nucleus and, if all goes as planned, the newly constituted cell proceeds to divide and develop like a naturally conceived embryo. This is how Dolly the sheep was produced.

In Altered Nuclear Transfer, the differentiated body cell nucleus or egg cytoplasm (or both) are first altered before the adult body cell nucleus is transferred into the oocyte. These alterations preclude the coordinated organization and developmental potential that are the defining nature of an embryonic organism. Nonetheless, the laboratory constructs produced by this process still have the capacity for a certain limited subset of growth sufficient to produce pluripotent stem cells.

6.6   Failures of Fertilization

There is natural precedent for such a project. In normal conception, fertilization signals the activation of the organizing principle for the self-development of the full human organism.

But without all of the essential elements – the necessary complement of chromosomes, proper epigenetic configuration and the cytoplasmic factors for gene expression – there can be no living whole, no organism, and no human embryo.

Recent evidence from infertility studies suggests that, in the natural reproductive process, incomplete or inadequate combinations of the necessary elements lead to many 'failures of fertilization.'

Some of these naturally occurring failures of fertilization may still proceed along partial trajectories of organic growth without being actual organisms. For example, certain grossly abnormal combinations of chromosomes (including haploid genomes, with only half the natural number of chromosomes) will form blastocyst-like structures but will not implant.

---

[2] In some techniques, the whole body cell is simply fused with the enucleated egg.

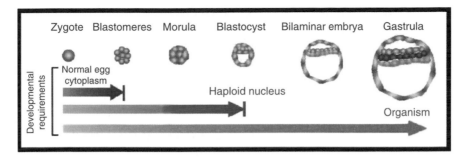

Fig. 6.2 With missing elements, organized growth and development may proceed in different ways and to different stages. BLUE: an egg cytoplasm without a nucleus will grow to the 8–16 cell stage. ORANGE: a fertilized egg with only half the normal number of chromosomes (haploid) will develop to the blastocyst stage. GREEN: to proceed forward as a fully integrated organism all the essential genetic and epigenetic elements must be present

(The diagram above is used by permission of Maureen Condic Ph.D. 9)

Even an egg without a nucleus, when artificially activated has the developmental power to divide to the eight-cell stage, yet clearly is not an embryo – or an organism at all. The mRNA for the protein synthesis that drives these early cell divisions is generated during the maturation of the egg and then activated after fertilization. Like a spinning top, the cells have a certain biological momentum that propels a partial and unorganized trajectory of development, but unlike an embryo, they are not adequately constituted to establish the dynamic molecular interactions that characterize a coherent and self-regulating organism.

Some of these aberrant products of fertilization lacking the qualities and characteristics of an organism appear to be capable of generating ES cells or their functional equivalent. Mature teratomas are benign tumors that generate all three primary embryonic cell types as well as more advanced cells and tissues, including partial limb and organ primordial – and sometimes hair, fingernails and even fully formed teeth. Yet these chaotic, disorganized, and nonfunctional masses are like a bag of jumbled puzzle parts lacking entirely the structural and dynamic character of organisms. Neither medical science nor the major religious traditions have ever considered these growths to be 'moral beings' worthy of protection, yet they produce embryonic stem cells.

These benign ovarian tumors appear to be derived by spontaneous development of activated eggs. The disorganized character of teratomas appears to arise, not from changes in the DNA sequence, but from genetic imprinting, an epigenetic modification that affects the pattern of gene expression (keeping some genes turned off and others on). In natural reproduction the sperm and egg have different, but complementary, patterns of imprinting, allowing a coordinated control of embryological development. When an egg is activated without a sperm, the trophectoderm (the outer layer in a natural embryo) and its lineages fail to develop properly. In the absence of the complementary genetic contribution of the male, the activated egg is simply inadequately constituted to direct the integrated development characteristic of human embryogenesis.

6.7   Systems Biology

This example points to another new dimension of our advancing knowledge. Through systems biology, we are beginning to recognize how even a small change of one gene can affect the entire balance of an enormous network of biochemical processes necessary to initiate and sustain the existence of a living being.

Systems biology offers us the view of an organism as a dynamic whole, an interactive web of interdependent processes that express emergent properties not apparent in the biochemical parts. Within this dynamic self-sustaining system is the very principle of life, the organizing information and coordinated coherence of a living being. With the full complement of coordinated parts, an organismal system subsumes and sustains the parts; it exerts a downward causation that binds and balances the parts into a patterned program of integrated growth and development. Partial organic subsystems (cells, tissues and organs) that are components of this larger whole, if separated or separately produced, may temporarily proceed forward in development. But without the coherent coordination and robust self-regulation of the full organism, they will ultimately become merely disorganized cellular growth.

The underlying idea of ANT is that small but precisely selected genetic or epigenetic alterations performed prior to nuclear transfer will allow the laboratory construction of cellular structures that are capable of producing pluripotent stem cells, but are not embryos. They are biologically (and therefore morally) equivalent not to embryos, but to teratomas and other fragmentary and unorganized growths.

6.8   Cdx2

Altered nuclear transfer is a broad concept with a range of possible approaches; there may be many ways this technique can be used to accomplish the same end.

One possibility involves the deletion or silencing of a gene necessary for the most fundamental level of coordinated development and organization. As described in January 2006 in the journal *Nature*, MIT stem cell biologists Rudolf Jaenisch and Alexander Meissner, through a series of mouse model experiments, have established the scientific feasibility of this approach. Using RNA interference, they were able to reversibly silence the functional expression of the gene Cdx2 in the somatic cell nucleus prior to its transfer into an enucleated egg (Meissner and Jaenisch 2006).

Without the expression of Cdx2, there is a fundamental failure of formation: the first differentiation into distinct cell lineages does not occur, the body axes do not form, and the basic body plan is never established. Yet from these laboratory constructs lacking the character of an organism, they procured fully functional pluripotent stem cells.

More recently, studies suggest that it may be possible to achieve the same results through the preemptive silencing of Cdx2 messenger RNA in the egg, in this case too before the act of nuclear transfer. University of Missouri biologist Michael Roberts reports that in natural mouse development maternally derived mRNA for Cdx2 is present in the egg and appears to be asymmetrically distributed in the first cell division after fertilization. The asymmetric distribution of Cdx2 mRNA may direct the cells at the two-cell stage to initiate distinct cell lineages. In mouse studies, by the selective silencing of Cdx2 in a single blastomere at the two-cell stage, Roberts was able to produce an unorganized mass composed exclusively of cells with the character of those of the ICM.[3]

Furthermore, in a recent presentation to the President's Council on Bioethics, Han Scholer, Director of Cell and Developmental Biology at the Max Planck Institute in Muenster, Germany, discussed similar mouse studies in which he silenced Cdx2 in the pronuclear stage, before division of the newly fertilized egg.[4] He described the product of this procedure as 'a single lineage tissue culture,' lacking the organization essential to be designated an embryo. Nonetheless, he was able to produce pluripotent stem cell lines from these cellular systems at an earlier stage (eight cells) and at a rate 50 percent higher than from natural embryos.

Scholer has studied the gene expression pattern of these Cdx2-deficient constructs and has documented their dramatic difference from the pattern associated with natural embryogenesis. By the eight cell stage, he found over 300 genes were up-regulated or down-regulated by a factor of three-fold or more over the natural pattern. He suggests that a similar result is likely from Cdx2 silencing in the oocyte with ANT.

This is the organic equivalent of a model airplane kit without the glue, you have parts but no capacity to form a coherent whole. In the absence of expression of

---

[3] Due to misrepresentations in some of the images, the original article (Deb, Kaushik, et al. "Cdx2 Gene Expression and Trophectoderm Lineage Specification in Mouse Embryos," Science, vol. 311, no. 5763 (Feb. 17, 2006) 992–996) has been retracted. However, the findings described in this paragraph have been replicated and appear to be as reported (personal communication with Michael Roberts). These findings concerning Cdx2 may be strain-specific in mice, and may not apply to primate development. Nonetheless, this example provides a useful description of the kind of alteration sought in ANT. As discussed above, ANT is a broad conceptual proposal and there could be many specific gene targets.

[4] Scholer, Hans, Testimony to the President's Council on Bioethics, Session 1, Stem Cell Research Update, November 16, 2006. http://www.bioethics.gov/transcripts/nov06/session1.html (last accessed November 29, 2006)

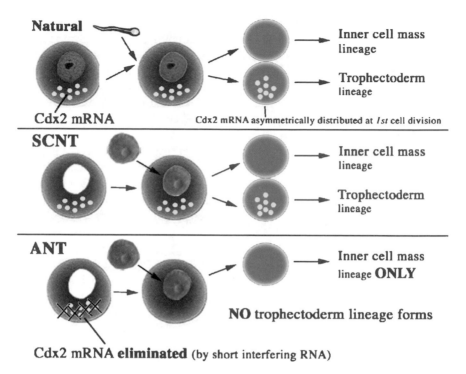

Fig. 6.3 With both natural fertilization and SCNT, the egg begins with the full allotment of Cdx2-mRNA necessary for growth and development. With ANT, the Cdx2-mRNA is removed by gene silencing using 'short interfering RNA' before the act of nuclear transfer is performed. When the altered nucleus of the somatic cell is transferred into the altered egg cytoplasm, no embryo is produced since Cdx2-mRNA is essential for natural growth and organization. Instead, just the lineage of the inner cell mass is produced, from which 'embryonic-like' stem cells may be obtained. Since the source of these stem cells is not rightly termed an embryo, they are called 'pluripotent stem cells'

Cdx2, as with a teratoma, the trophectoderm fails to grow and there is only partial and unorganized cellular process. But the gene Cdx2 does more than establish the cell lineage of the trophectoderm, it has been shown in mouse models to be essential for the early integration of organismal function. Without the trophectoderm the coordinated pattern of cell-cell interactions do not occur. These are crucial for further differentiation and development. Lacking one of the two essential cell types is the equivalent of trying to sing a duet with only one voice. The coordinated interactions that are essential for embryonic development are simply not possible. Nonetheless, something like an inner cell mass is produced from which functional embryonic stem cells can be extracted.

It is important to recognize that the improper development of the trophectoderm is not reasonably considered a defect within a part but rather a failure in the formation of the whole. An early embryo does not have parts in quite the same sense as an adult organism or even as a later-stage embryo just a few days or weeks later. Natural embryogenesis is, by definition, the period during which the whole, as the unified principle of growth, produces the parts. The differentiation of parts during early embryogenesis lays down the fundamental axes, body plan, and pattern of integrated organogenesis.

An embryo does not have a central integrating part like the brain; rather, the essential being is the whole being. At this stage, a critical 'deficiency' is more rightly considered an 'insufficiency,' not a defect *in* a being, but an inadequacy at such a fundamental level that it precludes the coordinated coherence and developmental potential that are the defining characteristics of an embryonic organism. In testimony to a US Senate subcommittee on stem cell research, Dr. Jaenisch stated: 'Because the ANT product lacks essential properties of the fertilized embryo, it is not justified to call it an "embryo." '[5]

Many scientists, moral philosophers and religious authorities (including some of the most conservative evangelical and Catholic leaders) have expressed strong encouragement for further exploration of this project. Of course additional animal studies, including some with non-human primates must precede any translation of these findings into practice with human cells.

6.9 Advantages of ANT

ANT, in its many variations, could provide a uniquely flexible tool and has many positive advantages that would help advance stem cell research.

- Unlike the use of embryos from IVF clinics, ANT would produce an unlimited range of genetic types for the study of disease, drug testing and possibly generation of therapeutically useful cells.
- By allowing controlled and reproducible experiments, ANT would provide a valuable tool for a wide range of research, including studies of gene expression, imprinting, and intercellular communication. For this research it will be essential to study the simplest cell-cell interactions, most specifically, those in the first few days of development. Once a morally acceptable form of ANT is established, it will provide an effective way to probe and explore the organic processes of these early dynamics of development.

---

[5] Jaenisch, Rudolf. 'Testimony of Rudolf Jaenisch, M.D., Hearing on 'An Alternative Method for Obtaining Embryonic Stem Cells', Committee on Appropriations, Subcommittee of Labor, Health and Human Services, Education', United States Senate Oct. 19, 2005.

- Furthermore, the basic research essential to establishing the ANT technique would advance our understanding of developmental biology and might serve as a bridge to transcendent technologies such as direct reprogramming of adult cells.
- Moreover, as a direct laboratory technique, ANT would unburden embryonic stem cell research from the additional ethical concerns of the 'left over' IVF embryos, including the attendant clinical and legal complexities in this realm of great personal and social sensitivity.

The one remaining link with IVF, the procurement of oocytes, is a subject of intense scientific research and there appear to be several prospects for obtaining eggs without the morally dubious and expensive hormonally induced super-ovulation of female patients. These include the use of eggs left over from IVF, the laboratory maturation of eggs cultured from ovaries obtained after surgical removal or from cadavers, and possibly the direct production of eggs from embryonic stem cells.

6.10   Conclusion

We are at a crucial moment in the progress of science and civilization. Advances in biology have delivered new powers with extraordinary potential for positive application in both basic research and clinical medicine. Yet, at the same time, these new possibilities challenge the most fundamental moral principles on which our society is based.

The English author G.K. Chesterton had a metaphor that may inform our current situation. Little boys are playing football on an island, but at the very edges of the field cliffs go down hundreds of feet to the waves crashing against the rocky shore. The boys are playing, but only in the middle 20 yards – no one wants to do a corner kick. Then someone comes and builds a sturdy fence right at the edges of the field: now they can play within the full field without fear of falling off the cliff.

Our current conflict is like this. If we can define with clarity and precision the moral boundaries we are trying to defend, we might open a wider arena of legitimate study without fear of the grave dangers posed by a breach of the basic moral principles that sustain our civilization.

The moral analysis that underlies the proposal for Altered Nuclear Transfer begins with a reaffirmation of the inviolability of human life across all of its stages from conception to natural death. By grounding this moral valuation in the continuity of organismal existence, we can then draw on the distinction between the full totipotency of a living being and the mere pluripotency of an organic subsystem with fragmentary cellular growth.

The conceptual shift essential for the practical acceptance of this proposal is based on an acknowledgment of the moral neutrality of laboratory constructs with only partial and unorganized developmental potential. As a society we have already accepted similar shifts when we overcame our initial reservations concerning the use of human parts separated from their organismic wholes as for example in blood transfusion and organ transplantation. This conceptual transition, however, will be more difficult because our natural moral sentiments equate the dynamic of growth and development with the powers and potentials of a living being. Nonetheless, as we enter the age of developmental biology we will come to understand that a biological artifact similar in character to a teratoma does not have the moral status of a human being, even though it can undergo the organic processes that produce human pluripotent stem cells.

As we enter the coming era of rapid advance in biotechnology, the search for 'alternative sources of human pluripotent stem cells' establishes a positive precedent for maintaining constructive ethical dialogue and encouraging creative use of our scientific knowledge. In recognizing the important values being defended by both sides of our difficult debate over embryonic stem cell research, this approach could at once, sustain social consensus and open hopeful prospects for scientific advance.

References

Eggan, Kevin, et al. (2005). "Nuclear reprogramming of somatic cells after fusion with human embryonic stem cells", *Science*, 26 Aug 2005; 309(5739):1369–1373. DOI:10.1126/Science.1116447.
Jaenisch, R., et al. (2006). *PNAS Online,* Week of 1/16/2006.
Joyce, Robert E. (1978). "Personhood and the conception event", *New Scholasticism* 52:97–109.
Meissner, Alexander and Jaenisch, Rudolf, (2006). "Generation of nuclear-transfer derived pluripotent ES cells from cloned *Cdx2*-deficient blastomeres", *Nature* 439:212–221.

Part II
Social and Political Perspectives

Chapter 7
An Intercultural Perspective on Human Embryonic Stem Cell Research[1]

LeRoy Walters

Abstract Four major policy options regarding human embryonic stem cell (hESC) research have been adopted by various nations around the world. These options can be described as the restrictive option, the permissive option, the moderate option, and the compromise option. Color-coded maps indicate the policies that have been adopted in seven world regions. In general, the worldwide trend since 2001 has been toward more permissive policies on hESC research. The United Nations also debated the questions of reproductive and research cloning between 2001 and 2005, reaching a compromise on a declaration in March 2005. Humans can be forgiven for not having yet reached firm ethical conclusions regarding in vitro embryos because such embryos only entered public consciousness in 1978 – with the birth of the first infant whose life was initiated through in vitro fertilization (IVF). Nonetheless, a broad international public consensus has emerged on the morality of IVF for reproductive purposes. From this consensus it is but a short step to the ethical acceptance of research involving IVF embryos no longer needed for reproductive purposes. The creation of embryos specifically for research purposes raises more difficult questions. The major eastern and western religious traditions have debated the abortion issue for centuries, and their discussions of this partially-analogous question may help to illuminate the ethical analysis of hESC research. More specifically, developmental views regarding our moral obligations toward human embryos and fetuses are quite compatible with an acceptance of research on human embryos in vitro.

Keywords Cloning, embryo, in vitro fertilization, public policy, stem cell

Joseph P. and Rose F. Kennedy Institute of Ethics and Georgetown University, Healy Building, Room 415, 37th and O Streets, N.W, Washington, DC 20057-1212.
e-mail: waltersl@georgetown.edu

[1] This essay seeks to update Walters (2004a).

L. Østnor (ed.), *Stem Cells, Human Embryos and Ethics: Interdisciplinary Perspectives.*
© Springer Science + Business Media B.V. 2008

Four major policy options regarding human embryonic stem cell (hESC) research have been adopted by various nations around the world and by the various states or regional governments of some nations. In choosing terms to describe these policy options, I have sought to be as neutral as possible. That is, these descriptors are intended to be non-prejudicial.

- The Restrictive Option: Prohibits human embryo research; does not explicitly permit research with existing human embryonic stem cell lines (shown in red on the following maps)
- The Permissive Option: Accepts the production of human embryos for research purposes through in vitro fertilization and/or nuclear transfer (cloning) (shown in green)
- The Moderate Option: Permits the derivation of new human embryonic stem cell lines but only through the use of remaining embryos from infertility clinics (shown in blue)
- The Compromise Option: Permits research with existing human embryonic stem cell lines but not the derivation of new stem cell lines through the destruction of human embryos (shown in yellow)

There is an important distinction to be made within each of these policy options, namely, the distinction between allowing or prohibiting certain kinds of research, on the one hand, and providing public funding for certain kinds of research, on the other. I will focus primary attention on what research is allowed or prohibited, with only passing references to what research is publicly funded. In countries where there is a substantial private sector, with minimal government regulation for that sector, the possibility of a substantial difference between what is permitted and what is publicly funded is more likely to arise.

7.1   Seven World Regions: An Overview

Europe

The map of Europe (Fig. 7.1) reveals substantial differences in public policies on hESC research. The countries adopting the Permissive Option are the United Kingdom, Belgium, Sweden, Finland, and Spain (shown in green). Countries in which the Restrictive Option is currently operative include Ireland, Austria, and Poland (shown in red). Perhaps the largest number of European nations have adopted the Moderate Option (shown in blue). Two large industrialized nations, Germany and Italy, permit the importation and use of human embryonic stem cells produced abroad, but not the production of such cells within the nation (shown in yellow). They have thus opted for the Compromise Option. In Germany, there is also a cutoff date, January 1, 2002, by which the stem cells must have been produced.

The funding policy of the European Union, under its 7th Framework Programme, seems to be the Compromise Option. No EU funds may be employed to produce human embryonic stem cells. During the previous funding cycle – the 6th Framework Programme – the EU policy seems to have more closely approximated the Moderate Option.

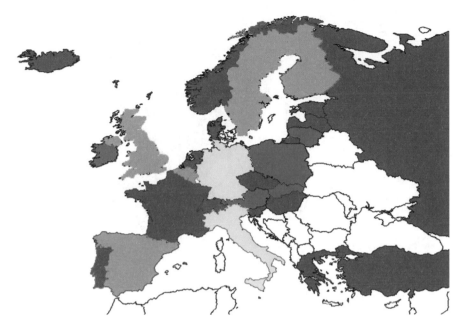

Fig. 7.1 Europe

The Middle East, the Persian Gulf, and Africa

In Israel (Fig. 7.2) the Permissive Option, at least with respect to nuclear transfer, has been a long-standing policy. Iranian scientists have succeeded in producing and publishing a human embryonic stem cell line. Iran seems to have adopted the Moderate Option.

In South Africa (Fig. 7.3), a bill that would permit nuclear transfer research using human somatic cells and egg cells has been introduced, but at last report it had not yet been enacted into law.

Asia

When one views the map of Asia (Fig. 7.4), one is immediately impressed by the fact that several large industrialized nations have adopted the Permissive Option. These nations include India, China, South Korea, and Japan. The smaller nation of Singapore, after extensive public debate, has also espoused the Permissive Option. Taiwan seems to have adopted the Moderate Option.

Australia and New Zealand

In Australia (Fig. 7.5) a major policy change is currently in progress. Through December 2006 the uniform national policy was the Moderate Option. However,

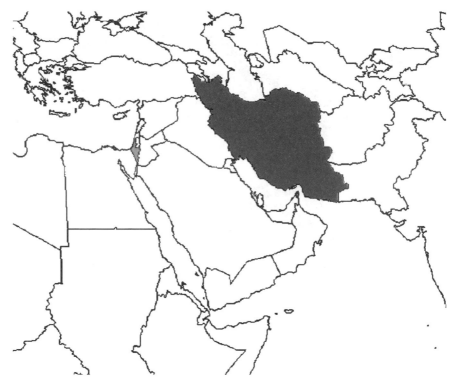

Fig. 7.2 The Middle East and the Persian Gulf

the Australian Parliament voted in December 2006 to make the Permissive Option national policy, effective six months after Royal Assent. The proposed new policy explicitly permits nuclear transfer for research purposes. The various states of Australia are now voting on whether to accept this new policy. As of July 2007, two states, Victoria and New South Wales, had done so.

New Zealand (Fig. 7.6) has adopted the Moderate Option. There is, however, ongoing public discussion about whether to permit nuclear transfer for research purposes, that is, about whether to adopt the Permissive Option.

South and North America

Brazil (Fig. 7.7) has adopted the Moderate Option as its public policy.

Canada (Fig. 7.8) has a uniform national policy and a national review committee for a human embryonic stem cell research conducted in Canada. It has adopted the Moderate Option.

There has been considerable public debate in Mexico about hESC research policy, and several proposals to place restrictions on the research have been

Fig. 7.3 Africa

considered by the Parliament. However, no bills on this subject seem to have been enacted into law.

The foregoing discussion summarizes what public policies were in place internationally in mid-year 2007. A more dynamic presentation would indicate that from 2002 to 2007 no fewer than 19 nations formally liberalized their hESC research policies – sometimes more than once. These nations are Australia, Belgium, Brazil, the Czech Republic, Denmark, France, Germany, Greece, Japan, the Netherlands, New Zealand, Norway, Portugal, Singapore, South Korea, Spain, Sweden, Switzerland, and the United Kingdom.

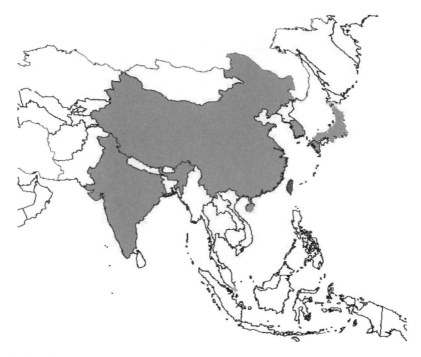

Fig. 7.4 Asia

7.2 The United Nations Debate about Human Cloning

In August 2001, the U.N. representatives of France and Germany brought forward a proposal that would have urged member nations to commit themselves through formal treaties to ban reproductive cloning. However, in February 2002, the United States delegation, supported by the non-voting representative of the Holy See (the Vatican), urged that research cloning should be included in the ban. This suggested change in the content of the U.N. resolution sparked a heated, multi-year debate, with the United Kingdom and several Islamic nations leading the opposition to the U.S. amendment.[2] A fatwa issued by a Sunni Egyptian scholar became an influential factor in this debate. The crucial section of this fatwa, written in 2003 by Prof. Dr. Ahmed Mohammad Al-Tayyeb of Al-Hazar University in Cairo, reads as follows:

> Now that we have discussed human cloning, let us discuss cloning in other fields, such as plant cloning or cloning in medical fields with the purpose of having transplantable body parts that can replace human parts that had been lost or have been malfunctioning. If cloning for this purpose is being undertaken – provided it has been sufficiently tested and has proved to be effective – then it may be considered as lawful and permissible. It may is encouraged by Islam which supports each scientific study conducted in the best interest of man whether on the moral or material aspect.

> Therefore, it can be concluded that: Human cloning is totally unlawful from the Sharia point of view. For it is beyond the pale of the Way that God, the Almighty, set down for man as His Viceroy on earth. It is beyond the pale of moral and social framework as set

---

[2] For a detailed account of the early stages of this debate, see Walters (2004b).

Fig. 7.5 Australia

Fig. 7.6 New Zealand

Fig. 7.7 South America

down in the Ever-Glorious Qur'an for man. Cloning parts of the human body so as to replace parts that ill and sick people have lost or as a therapeutic tool to treat some diseases, it would be considered as lawful. Moreover, cloning is also lawful for the purpose of increasing the productivity of plants or improving animal stock, provided that it neither involves a detrimental effect on the environment nor runs against the interest [of] God, the Almighty, set down for the whole universe: man, the flora and fauna and the inanimate objects.[3]

---

[3] Professor Al-Tayyeb kindly sent me a copy of this fatwa, in Arabic and English translation, on March 6, 2004.

Fig. 7.8 North America (outside the U.S)

What this text seems to say is that reproductive cloning if prohibited by Islamic law, but cloning for the purpose of treating disease is permissible.

Through the years 2002 through 2004 there was a virtual standoff at the U.N. on the cloning issue. By the end of 2004, the Costa Rican delegation had emerged as the leading advocate of a resolution that would have banned both reproductive cloning and nuclear transfer for research purposes. Costa Rica was strongly supported by the United States and by representatives of several developing countries, who feared that poor women from their nations might be exploited as oocyte providers if nuclear transfer research did in fact proceed. In the end a compromise was reached. There would be no agreement regarding an international convention, or treaty. Instead, there would be a declaration. The Italian delegation provided the text that was put to a final vote in March 2005. The declaration passed by a vote of 84 to 34, with 37 nations abstaining. The key text in the compromise Italian declaration (L.26) reads as follows:

> (b) Member states are called upon to prohibit all forms of human cloning inasmuch as they are incompatible with human dignity and the protection of human life [...]

Nations voting in favor of the resolution included Australia, Austria, Germany, Hungary, Ireland, Italy, Mexico, Poland, Portugal, Switzerland, and the United States. Nations voting against the resolution were Belgium, Brazil, Canada, China, Denmark, Finland, France, India, Japan, the Netherlands, New Zealand, Norway, the Republic of [South] Korea, Singapore, Spain, Sweden, and the United Kingdom. Among the nations abstaining were Egypt, Iran, Israel, and South Africa.

7.3 HESC Research Policies in the United States

The accompanying map (Fig. 7.9) reveals a patchwork of policies on hESC research in the United States. California was the first state to adopt the permissive option. It was followed by New Jersey, then several additional states with major biotechnology industries and/or large academic medical centers – Connecticut, Massachusetts, Maryland, Illinois, and Missouri. Again on the matter of state funding for the research California took the first step, gaining voter approval for a $3 billion bond issue that will provide a higher level of funding for the field – and for a wider variety of studies – than that provided by the National Institutes of Health.

U.S. federal policies for both funding and permission have undergone important changes within the past decade. During the second Clinton administration, 1996–2000, NIH planned to provide funding for research involving already-derived stem cells but not for the derivation of the cells. This policy expressly dissented from the slightly more liberal recommendations of the National Bioethics Advisory Commission, which would have funded derivation, as well, but only from embryos remaining after reproduction had been completed. President Bush narrowed the Clinton policy in August 2001, after an extended review of the issue. In his August 9, 2001, speech to the nation on hESC research, he restricted federal funding to stem cell lines that had already been derived *by a particular date*. At the time the NIH officials had informed the President that perhaps 60 to 70 such lines were available. In fact, almost six years later, only 21 approved lines are listed on the NIH Web site. On two occasions in 2007, both houses of the U.S. Congress voted to expand the number of hESC lines eligible for federal funding to include lines derived after

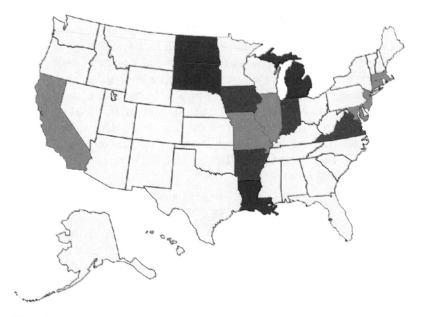

Fig. 7.9 HESC Research Policies in the United States

August 9, 2001, but in both cases President Bush vetoed the legislation. Thus, the current federal funding policy in the United States is the Compromise Option, with a time limit dating to 2001.

The more liberal requirements at the state level and the more conservative stipulations of federal funding policies have resulted in major inefficiencies in the way hESC research is conducted in the United States. In many cases academic research centers have had to divide laboratories or construct separate facilities and to duplicate equipment purchases, in order to ensure that federal funding for hESC research does not inadvertently 'cross over' into the use of stem cell lines that are not permitted by federal funding policies.

One other initiative of the Bush administration and several members of Congress merits at least brief mention. As noted above, in early 2002 the Bush administration began to advocate a position at the United Nations that would have committed signatory nations to ban both reproductive and research cloning. On the domestic front, President Bush and several members of Congress advocated legislation that would have placed a legal ban on research cloning, or nuclear transfer research using human cells. On two occasions between 2001 and the present this legislation was approved by the U.S. House of Representatives. However, the Senate refused to pass a similar bill, and research cloning continues to be permitted in states that have either expressly approved it or not expressly prohibited it. Had the federal anti-cloning bill been approved by the Senate and signed by the President, it would presumably have preempted the state legislation – for example, California legislation – that permitted research cloning.

7.4   Philosophical and Ethical Issues Surrounding hESC Research

We humans can be forgiven if we are as yet unsure how to regard early human embryos – either in vitro or in frozen storage. Those of us who follow the scientific literature carefully – and who are old enough to remember 1969 – have been aware that the earliest studies documenting the production of human embryos in vitro appeared just before 1970. However, for most laypeople, for most academics, and for most policymakers, the key date was 1978 – the year in which Louise Brown was born in the United Kingdom. Since that important date, perhaps two million infants have been born worldwide with the assistance of in vitro fertilization (IVF). At any given point in time, there are approximately 400,000 human embryos in frozen storage in infertility clinics in the United States.

Just 10 years ago, in 1997 the first mammalian clone was produced, again in the United Kingdom. Less than 10 years ago a University of Wisconsin research team succeeded in generating the first human embryonic stem cells. There is, as of July 2007, no clearly documented study in which human embryonic stem cells have been produced after nuclear transfer, or cloning.

How should we humans regard these thousands of early entities that we have deliberately produced and that we can freeze and store for years, if necessary – in the name of assisting individuals or couples to have genetically-related children? Philosophers, theologians, religious leaders, lawyers, policymakers, and opinion leaders have wrestled with these questions intensively, especially during the past 30 years.

From this international public discussion there has emerged what is clearly a majority ethical viewpoint, namely, that the fertilization of multiple human eggs in an effort to produce multiple early human embryos in the context of assisted reproduction is ethically acceptable. Further, there is a substantial majority that would also regard as ethically permissible the freezing and storage of embryos that are not transferred during a particular ovulatory cycle. The basis for accepting embryo freezing and storage is that this practice reduces the number of times that a woman needs to undergo hormonal stimulation (assuming that this technique is used) and egg retrieval.

The foregoing majority ethical viewpoint is not accepted by all. A Vatican Instruction from 1987 seemed to reject any union of sperm and egg that occurred outside a woman's body. Thus, gamete intrafallopian transfer (GIFT) seems not to have been condemned, whereas IVF was held to be ethically unacceptable. Disagreement with the majority view is also reflected in Costa Rican legislation that prohibits IVF and recent Italian legislation that requires that all early human embryos produced during a given ovulatory cycle must be transferred during that cycle. That is, no early embryos are to be frozen or discarded, and no preimplantation diagnosis is to be performed.

The metaphysical view of early embryos apparently presupposed by the majority viewpoint is that early embryos are a new type of entity that humans have never had to confront before. These entities are similar to sperm and egg cells because of their presence in vitro, yet they differ from gametes because they have a full complement of 46 chromosomes. In vitro embryos differ from embryos that have been transferred into the bodies of particular women in the sense that their future course is much less determinate and foreseeable. Such untransferred embryos are *candidates for further development*, but no more than that. Unless a woman steps forward to accept embryo transfer, and unless the progenitors of the embryo (who may include the woman) consent to the transfer, the embryo has no future. (In the future, if technologies for the extracorporeal development of human embryos from the blastocyst stage to maturity are developed, another option will be available. I will bracket this issue for the present discussion.)

The ethical question that arises when we view this new kind of entity is the following: What moral obligations do we have toward early human embryos? Again, there is clearly a majority ethical view, globally speaking. Within the context of a reproductive project, we have a moral obligation to protect IVF embryos from avoidable harm, to preserve their lives, and to foster their development. When that project has been completed, there exists no moral duty on the part of their progenitors, or others, to preserve the lives of IVF embryos. To express the same point in rights-language, no IVF embryo has a general right to life or a right to be transferred into the body of a woman, so that it may have the possibility of future development. The practical implication of this moral judgment is that (in most nations) individuals or couples are legally permitted to discard – that is, thaw and allow to die – embryos that are no longer needed for their reproductive projects. In some nations individuals and couples are legally required to discard IVF embryos after a specified time period.[4]

From this widespread practice it is but a short step, ethically speaking, to the proposal that in vitro embryos that are about to be discarded by individuals or couples who have completed their reproduction may be useful in research, if the progenitors of the embryos consent to such use. Some critics of this step have argued that the mere fact that an entity or a person is about to die (or to be allowed to die) should not justify performing research, especially research that destroys the entity, on the entity. However, it is possible to confine the application of the pro-research argument solely to in vitro embryos. It would not apply, for example, to proposals to perform research involving the fetus in utero in anticipation of abortion. In that case other considerations, such as damage to the fetus if the pregnant woman who is carrying it should change her decision about termination, come into play. In this case an in vitro embryo has no direct physical connection to anyone's body, and it is quite clear at a particular moment in time that the embryo will be allowed to die or used in research.

A moral threshold is crossed when one considers that possibility of creating embryos through IVF or (in the future) through nuclear transfer *specifically for research purposes*. The deliberate creation of IVF embryos for research purposes seems to some critics to constitute using the IVF embryos as means merely. Here a distinction should perhaps be drawn between nuclear transfer and simple IVF. With nuclear transfer there is currently the rationale that stem cells created through the use of this technique may have special characteristics in their nuclear DNA that are of particular interest to scientists and to people suffering from specific diseases, for example, amyotropic lateral sclerosis or juvenile-onset diabetes. The study of cells derived from such patients may provide insight into the early stages of pathological processes and, on the other hand, may suggest new approaches to treatment. At the same time, no scientist who is using nuclear transfer for research purposes would currently advocate embryo transfer into the body of a woman after nuclear transfer – that is, scientists who have spoken on this matter are uniformly opposed to reproductive cloning.

The most difficult case, unless one finds nuclear transfer problematic in principle, is the deliberate creation of embryos for research purposes through IVF. It may well be that remaining embryos from infertility clinics will be sufficient and that no recruitment of sperm and egg providers will be necessary in the research context. On the other hand, the freezing and thawing of IVF embryos reduces the probability of their development to the blastocyst stage, and the most symmetrical and best-developing IVF embryos will always be selected by infertility clinics for embryo transfer. Further, there may be a scientific rationale for recruiting sperm and egg providers with particular characteristics, for example, a recessive genetic

---

[4] Better methods of egg freezing may in the future help to alleviate the problem of remaining embryos. However, the success of those methods will need to be validated. Even with better egg freezing methods, there are likely to be remaining embryos.

trait in their genomes or a dominant gene for a late-onset genetic disorder like Huntington's disease.

The major moral argument for using IVF to create embryos for research purposes runs as follows. If one accepts the creation and the destruction of large numbers of IVF embryos for the legitimate familial goal of helping to produce genetically-related offspring, how much more should one be willing to accept this procedure for the sake of research directed toward the general good of humankind. The notion of general good includes short-term discoveries about the normal and abnormal development of human cells. In the long run, one hopes, the research will suggest new approaches to the prevention of disease or to therapy.

I should note that future technological developments could make the entire debate about human-embryo research irrelevant. Reports from three research groups in early June 2007 suggested that it may be possible to reprogram somatic skin cells – in this case, skin cells from mice – so that the reprogrammed cells will be functionally-equivalent to embryonic stem cells (Okita et al. 2007; Wernig et al. 2007; Maherali et al. 2007). A sobering aspect of this initial and welcome success is that one of the vectors employed in the reprogramming process also induced cancers in some of the modified cells (Okita et al. 2007) The prospect of being able, at some point in the future, to produce human embryonic stem cells without destroying human embryos is an attractive one. In the interim, however, researchers will quite reasonably want to continue research with methods that, at least in the case of IVF embryos, have a proven track record.

7.5 Religious Perspectives on hESC Research, Including Research Cloning

We should not expect major religious bodies to have worked out settled positions on issues that, in their current form, have only confronted them for about a decade. However, from publications and from position statements of religious leaders in Singapore, in particular, we can piece together an early picture.

Among the western religious traditions, the Jewish tradition has been highly supportive of hESC research from the beginning. There has been little, if any, criticism of using nuclear transfer as one of the methods for conducting the research. One important motive for the strong affirmation of this research by Jewish scholars and religious leaders is the tradition's emphasis of the moral duty to save human lives. The Jewish ethical tradition also considers our moral obligations to early human embryos – whether in vitro or in vivo – to be quite limited.

Perhaps more surprisingly, representatives of both the Sunni and Shi'ah traditions of Islam have affirmed the value and the ethical acceptability of hESC research. This positive response is not unanimous. However, a strong concern about human health and a developmental view about our moral obligations to human embryos and fetuses are both highly compatible with a positive response to hESC research.

Within the Christian tradition, Eastern Orthodox leaders, Roman Catholic leaders, and spokespeople for conservative Protestant groups have been critical of hESC research because it destroys – that is, kills – early human embryos.[5] Roman Catholic leaders and conservative Protestant spokespeople have also opposed using nuclear transfer to produce human embryonic stem cells, presumably on the grounds that post-nuclear-transfer embryos have the same moral rights as IVF embryos. A few Roman Catholic theologians have dissented from the position espoused by the leadership of their church. Among mainstream and liberal Protestants – including, for example Episcopalians and Presbyterians – there has been support for hESC research in church statements. Nuclear transfer for research purposes has received less attention than hESC following IVF.

It is perhaps perilous for a North American to discuss eastern religious traditions. In part because these questions are so new and in part because of the decentralized character of many eastern traditions, one hesitates to generalize. However, the Singapore Bioethics Advisory Committee (BAC) has been quite helpful in identifying possible tendencies in eastern religious thought – at least in one setting and culture. According to a 2002 report by the BAC, in Singapore representatives of Taoists and Sikhs opposed the destruction of human embryos for research purposes. The Hindu spokesman said, 'There is no non-acceptance to [using] these ES cells to protect human life and to advance life by curing diseases' (Singapore 2002: G-3-2). According to the spokesman for Singaporean Buddhists, the motives of the researchers were important for reaching a moral judgment. In the words of the Venerable Shi Ming Yi,

> Buddhism will look at [this research] seriously from the point of intention. If the intention of the research is to find cures specifically to human therapeutic[s]…, if the aim of the research is to help and benefit humankind, then we will deem the research as ethical. On the other hand, if the research is something just for the sake of doing [it] or simply to make money out of it, then we will feel it is unethical
>
> (Singapore 2002:G-3–33).

In sum, the Singaporean spokespeople for the Hindu and Buddhist religious traditions accepted research with IVF embryos if the research is done to promote human health. The spokespeople for the Taoist and Sikh traditions opposed the research. I can only express the hope that representatives of these venerable traditions from other nations and cultures will contribute their perspectives on this new and important set of issues.

7.6   The Quest for the Best Analogy to Our Issue

In the preceding parts of this essay I have sought to argue that the question of research with IVF embryos is at most 40 years old and that the issues surrounding clinical IVF have been debated for approximately the same length of time.

---

[5] In vitro therapy, in an effort to cure a disease in an embryo, followed by embryo transfer would be acceptable within the Roman Catholic moral tradition.

Thus, philosophers, theologians and others have had relatively little time in which to reach firm conclusions about our moral obligations to in vitro embryos – whether they are produced by means of IVF or nuclear transfer. I have also tried to present a plausible case for current practices in IVF clinics and for the ethical acceptability of research involving either remaining embryos or specially-created embryos.

In this final section I will look to the longer-term past in the hope of finding an analogous case that provides additional support for the viewpoint on hESC research that this essay espouses. That analogous case is the question of abortion. I will focus on western religious thought about this issue.

The only biblical text that remotely relates to the question of abortion is one involving the accidental causing of a miscarriage. The text is found in the Jewish Torah, at Exodus 21: 22–24.

> When men fight, and one of them pushes a pregnant woman and a miscarriage results, but no other damage ensues, the one responsible shall be fined according as the woman's husband shall exact from him, the payment to be based on reckoning.

> But if other damage ensues, the penalty shall be life for life, eye for eye, tooth for tooth, hand for hand, foot for foot, burn for burn, wound for wound, bruise for bruise
> (Jewish Publication Society 1962:136–137).

When this text was translated into Greek for the Septuagint, the translators replaced the distinction between 'damage' and 'no damage' with the distinction between an 'unformed fetus' and a 'formed fetus'. In subsequent thinking about pregnancy and moral obligations during pregnancy, the formed-unformed distinction became centrally important (Noonan 1970:6).

Within the Jewish and Muslim religious traditions a developmental view of the fetus in utero became – and remains – the majority view. Abortion until a certain stage of pregnancy is ethically permitted because the fetus in utero is relatively unformed and has not acquired certain characteristics (Tendler 2000; Sachedina 2000; Zoloth 2000; Osman 2004; Zoloth 2004).

In the early centuries of Christianity there was diversity of opinion on the question of abortion. In a Roman Empire where abortion was widely practiced, some Christian theologians argued that every abortion was a homicide (Noonan 1970:7–14). On the other hand, the 'formed-unformed' distinction came to prevail in the mainstream, or most authoritative, Christian theological and penitential traditions. Augustine presaged the predominant view when he argued that an unformed fetus had no soul and no sentience (Noonan 1970:15–16). His view was accepted by Thomas Aquinas and by most theologians through at least the 18th century (Noonan 1970:34–36). There is a nuance here that I do not want to obscure. Both the abortion of an unformed (that is, unensouled) fetus and of a formed (ensouled) fetus were considered to be *sins*. However, terminating the life of an unformed fetus was morally equivalent to the sin of contraception. In contrast, terminating the life of a formed fetus was considered to be (unjustified) homicide (Noonan 1970:15–18).

The predominant Christian view was increasingly called into question in the 18th and 19th centuries. Finally, in 1869, the authoritative Roman Catholic view came to be that it was morally safer to assume that ensoulment occurs at the time of fertilization.

Abortion at any stage of pregnancy was not, therefore, morally justified (Noonan 1970:39). This conclusion has been challenged by multiple Roman Catholic theologians since 1869 but remains the official teaching of the Roman Catholic Church (Noonan 1979:39–50; Farley 2000; Pellegrino 2000; Cahill 2004).

The conclusion that I draw from this historical survey is that, in the somewhat analogous case of abortion, the majority view in Judaism and Islam has been that our moral obligations to early embryos and fetuses can be overridden by other moral obligations. For most of Christian history the predominant view has been that abortion in the early stages of pregnancy more closely resembles contraception than homicide.

7.7 Summary

HESC research is a relatively novel issue for both ethics and public policy. The issue has been intensively debated in numerous nations and cultures. I have sought to argue that the production and discarding of embryos in clinical IVF is morally justifiable and that the use of in vitro embryos in research, whether that research is basic or applied, is also morally justifiable. Bioethics committees and commissions in multiple nations have debated the ethics of hESC research at length, and a majority of those groups have concluded that the research is morally justifiable, if there is appropriate public oversight. The maps of major world regions indicate that public policymakers are increasingly accepting the judgment that hESC is morally justifiable and are enacting policies that ensure that it is legally permitted. In many cases the policymakers are also voting to expend public funds to support the research.

There are two cautions that I would like to express in conclusion. The first caution is that it is too early to know how valuable this line of research will be. We also cannot be sure to what extent the destruction of early human embryos will continue to be an essential means for conducting hESC research. Second, there is strong competition among scientists, nations, and even states of the United States in the conduct of hESC research. Intellectual property rights have also emerged as an important arena of conflict.[6] In response to these incentives and potential conflicts, several nations have established centralized oversight bodies for hESC research. In my view, such oversight bodies help to foster the accountability of researchers, companies, and academic institutions. By informing both the public and policymakers about developments in research, these oversight bodies also promote the transparency of this relatively-new research field. In my view, such oversight committees should accompany hESC research during its early years – until a more substantial consensus is reached on both ethical guidelines and best practices.

---

[6] See the research guidelines of the International Society for Stem Cell Research published in February 2007 (ISSCR 2007).

Acknowledgments The following people provided general information for this essay: Cynthia Cohen (Georgetown University), Thomas Eich (Ruhr-University Bochum), Julia Finkel (Johns Hopkins University); Gail Javitt (Johns Hopkins University), Lori Knowles (University of Alberta), Alexandre Mauron (University of Geneva), and Erik Parens (the Hastings Center). The following people provided information for specific parts of the analysis: Ahmed Muhammed Al-Tayyeb (Islam); Robert Araujo (U N., Observer Mission of the Holy See); D. Balasubramanian (India); Zelina Ben-Gershon (Israel); Ole Johan Borge (Norway); Jan Carlstedt Duke (Sweden); Robin Alta Charo (multiple nations); Ole Döring (China); Mostafa Dolatyar (U.N., Mission of the Islamic Republic of Iran); Carlos Fernando Diaz (U.N., Mission of Costa Rica); B. M. Gandhi (India); Ahmad Hajihosseini (U.N., Observer Mission of the Organization of the Islamic Conference); William Hoffman (multiple nations); Alissa Johnson (state legislation, U.S); Gareth Jones (New Zealand); Phillan Joung (South Korea); Young-Mo Koo (South Korea); Line Matthiessen-Guyader (European Commission, Directorate General: Research); Jonathan Moreno (NAS guidelines); Michel Revel (Israel); Adam Thiam (Islamic Fiqh Academy, Saudi Arabia); Carolyn Willson (U.N., Mission of the United States); and Laurie Zoloth (Judaism, South Korea). I also thank postdoctoral fellows at the Children's Hospital in Boston.

References

Cahill, Lisa Sowle (2004). "Abortion: Religious Traditions: Roman Catholic Perspectives." In: Stephen G. Post (ed.). *Encyclopedia of Bioethics*. 3rd ed., New York: Macmillan Reference USA/Thomson/Gale, 2004, 31–35.

Farley, Margaret A. (2000). "Roman Catholic Views on Research Involving Human Embryonic Stem Cells." In: National Bioethics Advisory Commission. *Ethical Issues in Human Stem Cell Research*, Volume 3: *Religious Perspectives,* Rockville, MD: NBAC, June 2000, D-1 to D-5.

ISSCR (2007). International Society for Stem Cell Research. *Guidelines for the Conduct of Human Embryonic Stem Cell Research* (February 2007): available at the following URL: http://www.isscr.org/guidelines/ISSCRhESCguidelines2006.pdf(accessed July 18, 2007).

Jewish Publication Society of America (1962). *The Torah*. 2nd ed., New York: Jewish Publication Society, 1962, pp. 136–137.

Maherali, Nimet et al. (2007). "Directly Reprogrammed Fibroblasts Show Global Epigenetic Remodeling and Widespread Tissue Contribution", *Cell Stem Cell* 1, June 7, 2007, 55–70.

Noonan, John T., Jr. (1970). "An Almost Absolute Value in History." In: John T. Noonan, Jr. (ed.). *The Morality of Abortion: Legal and Historical Perspectives*. Cambridge, MA: Harvard University Press, 1970, 1–59.

Okita, Keisuke; Ichisaka, Tomoko; and Yamanak, Sinya (2007). "Generation of Germline-Competent Induced Pluripotent Stem Cells," *Nature* (online publication June 6, 2007).

Osman, Bakar (2004). "Abortion: Religious Traditions: Islamic Perspectives." In: Stephen G. Post (ed.). *Encyclopedia of Bioethics*. 3rd ed., New York: Macmillan Reference USA/Thomson/ Gale, 2004, 39–43.

Pellegrino, Edmund D. (2000). "Testimony." In: National Bioethics Advisory Commission. *Ethical Issues in Human Stem Cell Research*, Volume 3: *Religious Perspectives,* Rockville, MD: NBAC, June 2000, F-1 to F-5.

Sachedina, Abdulaziz (2000). "Islamic Perspectives on Research with Human Embryonic Stem Cells." In: National Bioethics Advisory Commission. *Ethical Issues in Human Stem Cell Research*, Volume 3: *Religious Perspectives,* Rockville, MD: NBAC, June 2000, G-1 to G-6.

Singapore (2002). Bioethics Advisory Committee. *Ethical, Legal and Social Issues in Human Stem Cell Research, Reproductive and Therapeutic Cloning*. Singapore: The Committee, June.

Tendler, Moshe Dovid (2000). "Stem Cell Research and Therapy: A Judeo-Biblical Perspective." In: National Bioethics Advisory Commission. *Ethical Issues in Human Stem Cell Research*, Volume 3: *Religious Perspectives,* Rockville, MD: NBAC, June 2000, H-1 to H-5.

Walters, LeRoy (2004a). "Human Embryonic Stem Cell Research: An Intercultural Perspective", *Kennedy Institute of Ethics Journal* 14(1): March 2004, 3–38.

Walters, LeRoy (2004b). "The United Nations and Human Cloning: A Debate on Hold", *Hastings Center Report* 34(1): January–February 2004, 5–6.

Wernig, Marius et al. (2007). "In Vitro Reprogramming of Fibroblasts into a Pluripotent ES-Cell-Like State", *Nature* (online publication June 6, 2007).

Zoloth, Laurie (2000). "The Ethics of the Eighth Day: Jewish Bioethics and Genetic Medicine: A Jewish Contribution to the Discourse." In: National Bioethics Advisory Commission. *Ethical Issues in Human Stem Cell Research*, Volume 3: *Religious Perspectives,* Rockville, MD: NBAC, June 2000, J-1 to J-26.

Zoloth, Laurie (2004). "Abortion: Religious Traditions: Jewish Perspectives." In: Stephen G. Post (ed.). *Encyclopedia of Bioethics*. 3rd ed., New York: Macmillan Reference USA/Thomson/Gale, 2004, 28–31.

Chapter 8
Human Embryo Research: The European Perspective

Egbert Schroten

Abstract In the European Union (EU), the subsidiarity principle applies in human embryo research: It is up to the member states to decide whether this kind of research is or is not allowed and on what conditions. The most important reason for the use of this subsidiarity principle is that there are big policy differences between the member states, based on fundamental differences in position as to the (moral) status of the human embryo. After a rough sketch of the situation in the EU the conclusion is that this EU perspective is not about to change soon, mainly because there is in practice no alternative. The differences between the member states are too big. However, the final part of this chapter is devoted to an attempt to start a discussion about what might be an alternative perspective. It is based on a distinction between two categories of human embryos in vitro (or in the freezer), namely embryos which are intended to be transferred into the womb and embryos which are, definitely or eventually, not. It is argued that this distinction is morally relevant in this sense that the last category does not have the same moral status as the first category. If accepted, this distinction could become the basis of a (EU) public policy concerning human embryo research.

Keywords Embryo research, European policy, European Union, status of the human embryo, public policy, ethics

8.1 Introduction

The invitation to take part in the Norwegian research project on the moral status of human embryos, with special regard to human embryonic stem cell (HESC) research and therapy, by giving some contribution from an European perspective, placed me in an ambivalent position. At one side I was thankful for the invitation

Ethics Institute, Utrecht University, P.O. Box 80 103, NL 3508 TC Utrecht.
e-mail: eschroten@freeler.nl

but at the other it raised some embarrassment, for the question is relevant: Is there really a European perspective and, if so, what is it?

In an attempt to summarize the position of the European Union (EU), I could quote Article 18 of the *Oviedo Convention on Human Rights and Biomedicine:*

1. Where the law allows research on embryos in vitro, it shall ensure adequate protection of the embryo.
2. The creation of human embryos for research purposes is prohibited.

On the basis of this summary, one could say that there is no EU position on whether embryo research is allowed. In the light of the first point one could compare the situation in the EU with the situation in Germany in the middle of the 16th century, about 40 years after the Reformation. In the Augsburg religious peace treaty (1555) the rule *Cuius regio, eius religio* ('whose region, his religion') was coined, i.e. the rulers could decide whether they accepted Lutheranism or Catholicism in their country, a decision which the inhabitants had to follow. In other words, in EU human embryo research, the subsidiarity principle applies: it is up to the member states to decide whether embryo research is or is not allowed. And, although it is clear from the wording of this first point that embryo research must be regulated by the member states, there is no explication of what 'adequate protection of the embryo' means. Thus, the interpretation of this expression is a matter of the member states as well.

Is it another sign of plurality in Europe? Yes, but that is a euphemism! It is not based on nice cultural varieties which should be cherished but it is, rather, a pragmatic or, if you like, political solution for a fundamental difference between the member states, based on fundamental differences in position about the moral status of the human embryo. Moreover, it should be kept in mind that several member states signed the Convention, but the only a few member states have ratified it. And some, perhaps, never will ratify because of point 2 of this Article 18. For, as we shall see, in some member states it is allowed, under strict conditions, to create human embryos for research purposes.

8.2   Sketching the Situation in the EU

Roughly speaking, the situation in the EU[1] comes down to this: In some countries, for instance the United Kingdom (UK), Belgium, Finland, and Sweden, embryo research is allowed under strict conditions. Other countries, for instance Denmark, France, Greece, Spain, and the Netherlands have regulations allowing the derivation of HESC from supernumerary embryos, again under strict conditions. In the Netherlands there

---

[1] I would like to thank here Katrin Hatzinger (Bureau Evangelische Kirche in Germany) in Brussels for some very useful information about the situation in the EU. Relevant information can be found in Matthiessen-Guyader Vol. I.

is a moratorium for creating embryos for research. It is not likely that it will be lifted soon. Countries like Estonia, Hungary, and Slovenia have no regulations, but allow some research on supernumerary embryos. Germany and Italy have regulations which restrict HESC research. These regulations mean that it is not allowed for scientists in these countries to derive new HESC, but it is allowed to import them. In Germany, these cells must have been derived before January, 2002. Austria, Ireland, Latvia and Poland have legislation prohibiting HESC.

Within the framework of this contribution, it is not feasible to elaborate on the regulations of every EU member states, but let me be a bit more specific in some cases. For instance Ireland:

Although there is no legislation dealing with embryo research there is a provision of the Irish Constitution from 1937 (amended in 1983), which provides as follows: 'The State acknowledges the right to life of the unborn and, with due regard to the equal right to life of the mother, guarantees in its laws to respect, and, as far as practicable, by its laws to defend and vindicate that right.' The right to life of the unborn, then, is equal to that of the mother. This leads to the question whether the fertilized egg and the embryo in vitro are seen as an unborn child? My impression is that the Irish authorities say 'yes' to that question, which implies that embryo research is not only against the law but even against the Constitution.

In the UK, embryo research is permitted under the Human Fertilization and Embryology Act (HFEA 1990) for any of five specified purposes, which are briefly:

- To promote advances in the treatment of infertility
- To increase knowledge about the causes of congenital disease
- To increase knowledge about the causes of miscarriage
- To develop more affective contraceptive techniques
- To develop methods for detecting gene or chromosome abnormalities in pre-implantation embryos

In 2001, regulations were made extending the purposes for which embryo research could be licensed:

- To increase knowledge about the development of embryos
- To increase knowledge about serious disease
- To enable such knowledge to be applied in developing treatments for serious disease

Finally, in the UK and in Belgium it is allowed, under strict conditions but nevertheless contrary to the Oviedo Convention, to *create* human embryos for the purpose of research.[2]

In Germany, the main legal framework is the Embryo Protection Law (Embryonenschutzgesetz). By this law the use of human embryos for research purposes is not permitted. However, because it does not exclude the import of HESC,

---

[2] At the time that this contribution was submitted (April 2007) the HFEA was in a process of revision

this kind of research is done, meanwhile with parliamentary approval (after intense discussions) in 2002, including, as has been said, that the HESC cells must have been derived before January, 2002.

Comparison between the UK and Germany shows that where embryo research is legislated, legislation either prohibits any kind of embryo research (Germany, Austria), or authorizes this research under specified conditions (UK, Sweden, Finland, Spain). On the other side, where embryo research is not legislated we see that in some member states this kind of research is nevertheless carried out (like in Estonia), whereas in other member states (like Portugal for instance) it is not (as far as I know!).

8.3   EGE, CDBI, CEC, and EC Research Policy

This may suffice to show what is the situation in the EU and, thus, what is the context of EU ethics committees and political institutions. Let us have a look at some of them: The European Group on Ethics in Science and New Technologies to the European Commission (EGE), the Steering Committee on Bioethics of the Council of Europe (CDBI) and EC research policy (the framework programs for research and development) as an attempt to 'translate' ethics into politics.

The EGE issued four Opinions which refer to embryo research:

- Ethical Aspects of Cloning Techniques (1997)
- Ethical Aspects of Research Involving the Use of Human Embryo in the Context of the 5th Framework Programme (1998)
- Ethical Aspects of Human Stem Cell Research and Use (2000)
- Ethical Aspects of Patenting Inventions Involving Human Stem Cells (2002)

More or less, all EGE Opinions have the same structure: They first give a sketch of the legal and ethical framework, then they offer a short analysis of the problem and, finally, the proper opinion is formulated in the form of numbered paragraphs.

In the Opinion on Cloning Techniques, the situation in the EU is mirrored in one of the paragraphs (2.9): 'Taking into account the serious ethical controversies surrounding human embryo research: for those countries in which non-therapeutic research on human embryos is allowed under strict license, a research project involving nuclear substitution should have the objective either to throw light on the cause of human disease or to contribute to the alleviation of suffering, and should not include replacement of the manipulated embryos in a uterus.'

The same message is conveyed in the Opinion on Stem Cell research. Paragraph 2.3 reads:

Pluralism is characteristic of the European Union, mirroring the richness of its tradition and adding a need for mutual respect and tolerance. Respect for the different philosophical, moral or legal approaches and for diverse cultures is implicit in the **ethical dimension of building a democratic European society** (bold by EGE).

And the press release adds that this is a '… reminder that it is for each Member State to legislate on the derivation of stem cells from human embryos', a clear example that in this area the subsidiarity principle applies.

The EGE itself, however, does not seem to be against embryo research. Although it rejects the creation of embryos for the sole purpose of research and it pleads for more research on stem cells derived from other sources than the human embryo (adult stem cells, stem cells from umbilical cord blood and fetal tissues), it does not reject in principle the creation of embryos by somatic cell nuclear transfer (2.7). Moreover, the Group pleads for a specific Community research budget for stem cell research 'based on alternative sources' and for integrating it in the Framework Programme of research, but in this context spare embryos are mentioned as well (2.8). And in paragraph 2.5 we can read that when embryo research

> […] is allowed, with the purpose of improving treatment for infertility, **it is hard to see any specific argument which would prohibit extending the scope of such research** in order to develop new treatments to cure severe diseases or injuries (bold by EGE).

It goes without saying that the EGE recommends that an ethical assessment of research on stem cells financed by the EC has to be carried out both before the launching of a project and in monitoring its implementation.

And that is what you see happening in EC research politics and, more specifically, in the 6th and 7th Framework Programmes (FP) of the European Community for research, technological development and demonstration activities. Like the 6th FP, the current 7th FP (Article 6) explicitly forbids EU funding for research that involves human reproductive cloning, the creation of human embryos for research and research that would change the genetic heritage (germ line intervention). However, HESC research on existing stem cell lines may obtain funding. As a difference from the 6th FP, however, the EC commits itself not to promote research projects which imply the destruction of human embryos. HESC projects are examined on a case by case basis by independent scientific and ethical experts and they will not be funded in a member state where HESC research is forbidden. Each project has to show a favorable opinion from a local or national ethics committee from each country where HESC research will be carried out. It means that each project with a component of HESC research submitted for funding at EU level will be scrutinized by at least two ethical reviews, one at a national level and one at EU level, but it can be more if the HESC research will be performed in more than one country. Those proposals that pass the scientific evaluation and the ethical review are submitted for opinion to member states, meeting as a regulatory committee. No project is funded that does not receive a favorable opinion from that committee. Priority is always given to research on adult stem cells, which pose less ethical problems. In short, HESC research can be funded but under very strict conditions.[3]

---

[3] In March 2007 the EC agreed funding for the creation of a EU registry for HESC lines (www. cmrb.eu, MEMO/07/122).

There is another EU institution which is involved in reflecting on ethical questions concerning research on human embryos. It is the Council of Europe and more in particular the CDBI. In 1992, its predecessor, the Ad Hoc Committee of Experts on Bioethics (CAHBI), began its work to develop a framework convention, setting out the standards for the protection of the human person in the context of the biomedical sciences. The result was the *Convention for the Protection of Human Rights and Dignity of the Human Being with regard to the Application of Biology and Medicine*, the so called 'Bioethics Convention' or Oviedo Convention, which was opened for signature in April 1997 and from which Article 18 has been quoted in the beginning of this contribution. The need to undertake more reflection on questions concerning the protection of the embryo in vitro and the use of medically assisted procreation lead to the setting up in 1995 of a Working Party on the Protection on the Human Embryo and Fetus (CDBI-CO-GT3).

A publication of this Working Party, issued in 2003 under the title 'The Protection of the Human Embryo In Vitro', offers us an overview of current positions found in Europe regarding this topic. It shows that the word 'pluralism', mentioned in EGE Opinions, is a euphemism in this context, because it is caused by a fundamental philosophical controversy about the status of the human embryo. Four main moral positions are identified, two opposite positions and two 'gradualist' positions:

- In the first case, a fertilized egg is regarded as a (future) human being. Therefore, a fertilized egg or an embryo has an inviolable value (like all human beings) and a right to life.
- The opposite position is that an embryo is considered to be a lump of cells that have little or no moral value. Hence, it is not considered to need any particular protection, let alone that it would be regarded as having a right to life.
- Holders of the 'gradualist' positions underline that the fertilized egg is gradually developing into a human being. The embryo is considered to have significant, but not absolute value. The first position is that, as development is a continuous process, entitlement to rights and protection increases progressively throughout development, with full protection and rights at the time of viability.
- The other 'gradualist' position holds that full protection and rights are only achieved at birth. It means for instance that holders of this position may find abortion acceptable at a later stage of pregnancy than holders of the first 'gradualist' position would.

It would lead to far to give a summary of the whole Report of the Working Party. Its conclusion is that there is '[…] a broad consensus on the need for the protection of the embryo in vitro. However, the definition of the status of the embryo remains an area where fundamental differences are encountered, based on strong arguments'.

Before making some closing remarks, I want to give also some information on another European institution, the Conference of European Churches, an ecumenical organization which links in fellowship some 125 Anglican, Baptist, Lutheran, Methodist, Old Catholic, Orthodox, Pentecostal and Reformed churches and several associated organizations in all the countries on the European continent. It has a

Working Group on Bioethics which has been active in the debate on embryo research.[4] It produced a position paper on human and animal cloning in 1998. In 2000, a discussion paper has been issued on therapeutic uses of cloning and HESC, inviting wider debate. In 2002, the Working Group gave also an opinion on stem cell patenting to the EGE. The 2000 discussion document reflects a diversity of views which exist among member churches of CEC on the status of the human embryo and HESC research. This document has got an update in 2005 in the light of developments in research in the last years. This update also draws attention to concerns that expensive HESC research may be a luxury of Northern lifestyle and asks how far this research is justified while countries promoting it still have not fulfilled their promises of the percent of their GDP they dedicate to aid and support for healthcare in the developing world.

These documents show that there is no unanimity in the churches. In the 2005 Update, just mentioned, we read:

> The prospect of using stem cells to provide replacement cells to treat a wide range of otherwise incurable degenerative diseases is a compassionate aim with which most agree. The primary ethical controversy is whether the human embryo can be used as a source for these cells. [...] Among our member churches there are many for whom all research on embryos which causes their destruction is completely unacceptable, as a matter of fundamental principle. [...] The position is 'under no circumstances'. Only adult or cord blood stem cell research is permissible. Many of our churches do not, however, share this view and consider that the status of the human embryo increases with development, and would allow embryo research under particular circumstances. [...] This is a 'yes, provided ...' position. Others argue, on the contrary, that to use embryos merely as a source of cells is too instrumental [...] (and) would therefore object to the creation of embryos for stem cell research, but might reluctantly agree to use of surplus embryos from IVF treatment, given that these would normally be destroyed. This position is 'No, unless ...'.

Coming back to my embarrassment in the beginning of my paper: Is there an EU perspective? Can we, in the light of what has been said, stick to the *Cuius regio, eius religio* characterization? I think we can and I would be surprised if this situation would change soon.

In other words, in Europe there is, on this point, at least for the time being, a subsidiarity principle, because there is no alternative. The differences between the member states are too big.

8.4 An Alternative Perspective?

However, it is interesting to think and discuss about what could be a viable alternative. That is why I want to offer my own position,[5] with no more pretension than that it might be a contribution to the discussion. I want to take my starting point in IVF.

---

[4] For more information csc@cec-kek.fr

[5] See also my contribution in McLaren et al. (2002), 87–102 (Is human cloning inherently wrong?)

We have to realize that IVF makes an enormous difference in comparison with the past. Until the seventies of the last century a human embryo was an embryo in the womb. Until then, everything that has been said about the moral status of the embryo (for instance in the discussion on abortion) was said about the embryo in the womb.

But, because of IVF and cryopreservation, whether we like it or not, we have, by implication, two categories of (human) embryos: embryos in the womb and embryos in vitro. And, to use a useful distinction made by the Working Party of the CDBI, these embryos in vitro can be embryos in a parental project and embryos (for some reason or another) not in a parental project. In other words, we have embryos which hopefully become children and embryos that will not become children. IVF and cryopreservation, then, made it possible to take an embryo outside a parental project. From an ethical point of view, this is a new situation.

The point I want to make is that, in my view, there is a moral difference between these two categories of embryos. Although they share human genetic inheritance, embryos in vitro, outside a parental project, will not become children, which means that they need not to be treated as future children (or future human beings). It is interesting to place this against the background of traditional reasoning about the moral status of the human embryo. It can be summarized by a sentence, derived from Tertullian,[6] a theologian/philosopher in the second century: *Homo est et qui est futurus* (a future human being is a human being too). This way of reasoning became problematic after the introduction of IVF, which created, as we have seen, the possibility to have embryos which are not destined to be 'future human beings'. According to me, their moral status is lower than the status of embryos within a parental project or in the womb, which have to be treated as future children.

An often used counterargument against this position is what I would call the 'ontological status' argument. The attempt to make a distinction in the moral status of the human embryo, relying on the destination of the embryos, does not take into account the ontological status of the embryo. Since it shares the human genetic heritage, an embryo is, after the process of fertilization, a potential human being. To avoid misunderstandings, I wouldn't deny that, of course. It is the very reason for me to acknowledge a special moral status for surplus embryos. However, I don't think that it is a strong argument against my position that the destination of embryos is morally relevant and should be taken into account in ethics and in policy making. Let me explain that as follows: The potentiality of the human embryo to become a human being is certainly there, but it is a very weak potentiality, if I may say so, because the actualisation of it is heavily dependent on external factors. In other words, it depends on conditions which do not belong to the potentiality or the 'ontological status' of the embryo itself. These conditions include: (The decision to) transfer into a womb, nidation, and a healthy pregnancy. If these conditions are not met, an embryo will never become a human being in the normal sense of the word. Quite the contrary, if a (human) embryo is left to its own inherent potentiality alone, it will die. In other words, sharing human genetic heritage is a necessary but

---

[6] *Apologeticum* IX,8

not a sufficient condition for becoming a human being. And since surplus embryos do not meet these conditions they will not become actual human beings and should, therefore, not be treated as such.

Or should they? In the wake of the ontological status argument there is the right to life argument. Every (any?) human embryo (potential human being) has the right to life (to become an actual human being), which means that we are obliged to bring them into the conditions as mentioned above (transfer into the womb, nidation and pregnancy). Let us have a closer look to this argument, for I don't think it is very convincing. And let us, for the sake of argument, avoid the difficulties of 'rights language' in the context of embryos (in casu fertilized eggs). The argument is, then, the (moral) claim that any human embryo should always have a fair chance to become a human being. The question is: Why? And perhaps also: How? This is an open moral question. For, if we would argue from the ontological status of the human embryo, it would be in my view a *petitio principii* (begging the question). If it is indeed an open moral question, it wouldn't be difficult to launch a *reductio ad absurdum* (refutation by absurd consequences) argument here, for instance in view of the fact (!) that, in nature, probably more than 60% of fertilized eggs are lost, or in view of the fact that there are presumably thousands and thousands of surplus embryos in Europe (not to mention the whole world!), many of which are biologically not in a good shape or damaged when they are defrosted.

Perhaps one should mention, in this context, a religious argument as well: The human embryo is God's creation. In this contribution I shall limit myself to two remarks. (1) The theological question is: What does that mean in practice? 'Creation' as a concept is not restricted to human embryos. It is a very wide concept and it is difficult to make it operational in ethics, partly because this concept is not clear, partly because it is applied in various ways, at least in Christian tradition. If it is taken to mean that we should respect human embryos, it is not of great help here. At least, the next question would be: How? In what way? (2) However, in Christian tradition (and as far as I know in Jewish and Muslim tradition), special attention is paid to the creation of the soul. In the main stream of this tradition the (rational) soul is created 40 days (for female embryos 90 days) after conception. Influence of Aristotle is unmistakable, but I shall not elaborate on that. What I want to say here is not that we should go back again to an embryology of the old days, but that the main stream of Western religious tradition purports that the soul is not created immediately after conception, which implies, as far as I can see, that human dignity is not to be attributed immediately after conception. Of course, we have left Aristotelian and Scholastic embryology behind us, but that does not imply that the question if and why we should attribute human dignity to a surplus embryo outside a parental project has been solved.

In this contribution, I have focussed on supernumerary embryos. But we are all aware of another issue in this context, namely the issue of culturing embryos specifically for research. Although here, from the beginning, the intention is to use the embryos for research and not to transfer them into the womb, which makes this procedure even more morally problematic, I would nevertheless claim that my position holds true in this context as well. These cultured embryos, too, will belong to the same category as surplus embryos, outside the parental project. This would mean, for instance, that although one could speak in this context of an instrumentalization of human life, it is not a question of instrumentalization of human beings in the normal sense of the word.

In short, then, what I want to say is that the distinction between human embryos on the basis of their destination is morally relevant. This claim is, as far as I can see, not refuted by arguments on the basis of the ontological status of the embryo, or on the basis of an embryo's right to life, or on the basis of the main stream in Christian tradition. So, being a Christian theologian, I want to underline that, as far as I can see, my claim may also stay upright in Christian ethics. Science and technology (in casu IVF and cryopreservation), whether we like it or not, lead us into new situations where we have to rethink traditional moral principles and values and where we have to make new distinctions. Of course, we should be critical but we should not bury our heads in the sand and refuse to face facts. Ethics is a dynamic undertaking and new situations cannot always be met by old answers.

My conclusion is that supernumerary embryos are to be morally distinguished from embryos which are intended to be transferred into an uterus. This distinction could become the basis of a (EU) public policy concerning the use of human embryos for research. However, since they share human genetic heritage they should be taken as a special category, which only under strict conditions could be used for research. These conditions would include for instance that there are no alternatives and that the research aims would be substantial and morally acceptable. These and other conditions would underline the special status of the human surplus embryo and in these terms the expression 'adequate protection', as it is used in the European Bioethics Convention, should be implemented.

References

European Group on Ethics in Sciences and New Technologies to the European Commission (EGE) Opinion nr. 9 (1997). "Ethical Aspects of Cloning Techniques", European Commission.
EGE Opinion nr. 12 (1998). "Ethical Aspects of Research Involving the Use of Human Embryo in the Context of the 5th Framework Programme", European Commission.
EGE Opinion nr. 15 (2000). "Ethical Aspects of Human Stem Cell Research and Use", European Commission (Revised version 2002).
EGE Opinion nr. 16 (2002). "Ethical Aspects of Patenting Inventions Involving Human Stem Cells", European Commission.
(The EU Publications Office website: http://publications.eu.int/)
McLaren A et al. (2002). "Ethical Eye: Cloning", Council of Europe Publishing, Strasbourg.
Matthiessen-Guyader (2004). "Survey on opinions from National ethics Committees or similar bodies, public debate and national legislation in relation to human embryonic stem cell research and use", Vol. I in *EU member states*, European Commission.
Oviedo Convention for the protection of Human Rights and dignity of the human being with regard to the application of biology and biomedicine (1997). Convention on Human Rights and Biomedicine. Opening for signature: April 4, 1997
Tertullian, Quintus Septimius Florens [197] (1947). *Apologeticum*. Lindeboom et al. (ed.) Amsterdam.

Chapter 9
Stem Cells, Pluralism and Moral Empathy

Theo A. Boer

We needn't lose sight of how much moral agreement is lurking on the background, nor need we interpret this agreement as a matter of brute, blind luck.

Susan Wolf

Abstract In discussions about the morality of Human Embryonic Stem Cell Research, the focus is often on the differences. In this essay, two points are made. First, it is argued that different standpoints do not necessarily imply that altogether different values are held. Rather, shared values, including the intrinsic value of embryos (persons or not) *and* the value of developing medical therapies, may conflict and are *weighed* differently. Secondly, since we tend to forget or downplay values which we override, it is argued that we need the moral virtue of empathy. Empathy enables us to see overridden values in our own position, it fosters understanding of the weighing made by our opponents, and it stimulates the search for alternatives to Human ES-cell research which respect all values involved.

Keywords Stem cells, embryos, pluralism, relativism, ethics

9.1 Introduction

In discussions about the use of human embryos for stem cell research (Human ES-cell research) there is often a deeply felt disagreement as to what we may or may not do with human embryos. Participants in the debate try to defend their case and to convince others. In many cases such attempts are fruitless and do not bring the parties any closer. Hence, opponents and advocates of Human ES-cell research get

Protestant Theological University, P.O. Box 80.105, NL-3508 TC Utrecht, The Netherlands.
e-mail: taboer@pthu.nl

L. Østnor (ed.), *Stem Cells, Human Embryos and Ethics: Interdisciplinary Perspectives.*   121
© Springer Science + Business Media B.V. 2008

entrenched in fixed positions. Here, as in other burning issues such as euthanasia, or resorting to violence, the parties sometimes seem to represent different moral universes. It comes close at hand to conclude that morality is a matter of unbridgeable differences between subjective moral positions and that moral disputes are to be solved by procedural means only. Why discuss any further?

The first purpose of this essay is to challenge the view that deep and lasting differences are the result of mere subjective preferences. Moral plurality may better be explained as an expression of *shared* moral values which transcend these preferences. The most important hypothesis is that when parties advocate different actions and policies, this neither means that morality is a matter of mere subjectivity, nor that in such cases only one party is necessarily right and the other wrong. Perhaps both are right, or partly right. If this is correct, a debate is not only possible and meaningful, but also necessary, even if this does not always guarantee rapprochement. In the second part of this paper (starting in Section 9.5) I argue that we need a mental capacity to make such pluralistic thinking possible in practice: moral empathy.[1]

9.2   Long Live Pluralism?

If there is lasting and deep disagreement about moral matters we often use the terms, 'plurality' and 'pluralism.' Although the terms are sometimes used as synonyms, there is an important difference between them: whereas 'plurality' merely points to the existence of multiple views and normally carries no evaluative connotations, 'pluralism,' like most 'isms,' does. The Dutch ethicist Frits de Lange describes pluralism as a 'manner of thought which accounts for the existence of differences and which knows how to appreciate them' (De Lange 1995). Pluralism, he argues, is a 'normative scheme of interpretation' which does not strive to reconcile opposing views at all costs. Between competing claims, there need not exist a relationship of exclusion or hierarchy.

> Pluralism is a theory of resistance against every form of monism, both religious and secular. Pluralists in ethics share the conviction that the plurality of values to which people adhere should not be considered as a factor of loss but rather as an asset to the benefit of the good life, even if these values are irreconcilable. Ethical pluralism always speaks about morality in plural: there are more morals and this is a good thing to be [...] Therefore: long live pluralism! A non rigoristic morality without *God's eye point of view*, without the total transparency of reason, but still with a minimal basis, a solid bottom line, a basic morality with two floors: a pluralistic upper store, on which a thousand private flowers may bloom, and a public basic morality which cannot be compromised.[2]

---

[1] The author wishes to thank the international project group, 'The Moral Status of Human Embryos with Special Regard to Stem Cell Research and Therapy' at Oslo University and the MF Norwegian School of Theology for valuable comments on an earlier draft of this paper.

[2] De Lange 2003:45. De Lange, who here uses John Kekes (1993:11), articulates this basic morality in the form of three principles: integrity, respect, and safety.

The British philosopher of religion Brian Hebblethwaite shares this positive view on plurality. He admits that plurality is not always a feast, especially when it is caused by scarcity, irreconcilable interests, and deficiencies in moral knowledge. But his main arguments for embracing plurality are stated more positively: each person has a unique individuality; different people have different personal vocations in life; if we defend the possibility of supererogation, this implies that more than one choice can be called 'morally right'; different circumstances make for different modes of interpersonal and community life, and cultural differences yield different forms of excellence; and, finally, different stages in personal growth and historical development may yield penultimate forms of goodness. The quest for a perfect and united morality not only flies in the face of how the world 'works,' but also yields a morality in which the best is made the enemy of the good: if we always opt for the best without being prepared to settle for the second-best, we may end up empty handed (Hebblethwaite 1997:56). And even if full perfection *would* be reached, there would still be a plurality of goods: God's goodness is in itself differentiated and brings forth a myriad of finite, contingent, human goodness [65]. Hence, Hebblethwaite prefers to refer to plurality as 'varieties of goodness.'

There is undeniably an intuitive appeal in these arguments for a positive attitude towards plurality. After all, new and unexpected expressions of goodness are often a reason for joy and amazement. Moreover, the views represented by De Lange and Hebblethwaite are coherent with the view that morality is rooted in an objective reality which surpasses subjectivistic convictions. The recognition that there may be more than just one moral good involved is one of the keys to depolarize a complex moral debate.

Pluralistic theories thus offer promising perspectives, but before we can explore their relevance and value for the debate on Human ES-cell research, we need to be able to see the problems of pluralism. Competing claims and irreconcilable values present not only blessings but heartbreaking dilemmas and persistent questions as well. What is to count as a 'good'? Do *all* well-considered moral convictions represent a 'good'? Which good should prevail in case of a conflict? Which part of morality is the field of the thousand flowers and which is the part that is nonnegotiable? And if differences in opinion continue to be deep and divisive, how are we to make decisions in the field of public policy?

Before these questions can be addressed, a more basic question needs to be considered: what does the existence of moral plurality tell us about the plausibility of the different positions? Can people make different claims and be right at the same time? Or are questions of plausibility and rightness futile here, because pluralism implies that morality is a matter of subjective choices? Suppose A finds purely instrumental use of embryos always wrong whereas B considers the prospect of medical breakthroughs a justification of their instrumental use. The first possibility is that either A or B is right. This may be convenient for both, since that implies that their views are not subjective expressions of individual preferences. Meanwhile, of course, both assume that their own opinion is the right one and the other is wrong. The alternative would be that perhaps *both* are right or partly right. But is this possible while still affirming morality's objectivity? Can we argue that opposed views may both be called 'true' without ending in relativism and even subjectivism?

9.3   Two Levels of Pluralism

In an instructive article in *Ethics*, entitled, 'Two Levels of Pluralism,' the American moral philosopher Susan Wolf elaborates two concepts of the term pluralism: 'pluralism without relativism' and 'relativism without subjectivism' (Wolf 1992). These concepts have differences as well as similarities. Both keep clear of the extremes of subjectivism and absolutism.

9.3.1   *'Pluralism without Relativism'*

The first kind of pluralism is 'pluralism without relativism.' According to Wolf, one may be a pluralist without embracing relativism (i.e., the view that moral truth depends on moral contexts). Even those who reject a context-bound morality may sometimes have two or more mutually exclusive options to choose from. Some rigoristic versions of Utilitarianism, Kantianism, and Divine Command theories do not know this possibility, since they all have one or more clear and independent principles which tell us what to do in the case of a dilemma. The pluralist does not always have such a principle. Wolf quotes Bernard Gert, who offers an analogy between the question, 'what is the best policy regarding euthanasia?' and the question, 'who is the best hitter in the major leagues?' The answer to the latter question depends on which quality, or combination of qualities, is taken into consideration: is it the number of home runs of a player? The highest batting average? The number of RBI's ('Runs Batted In')? Despite all these open questions, however, this much is evident: certain players do *not* qualify as best player. Undecidedness has nothing to do with subjectivism (Wolf 1992:790; Gert 1988:54). Too much attention to the problems in establishing the right decision criteria tends to divert our attention from the shared, objective basis: 'We needn't lose sight of how much moral agreement is lurking on the background, nor need we interpret this agreement as a matter of brute, blind luck' (Wolf 1992:791).

Pluralism on this level occurs inside one person or within one normative system. This explains not only the existence of moral dilemmas for individuals, or of Gordian knots within a moral community, but also the occurrence of moral conflicts within a normative theory, worldview, or religion. This kind of pluralism keeps clear from subjectivism and needs in principle not to be relativistic. In what follows, I will call this kind of pluralism 'value pluralism.'

9.3.2   *'Relativism without Subjectivism'*

Next, Wolf identifies a pluralism which is prepared to accept relativism, but which still keeps clear from subjectivism. This 'relativism without subjectivism' entails the view that several plausible moral systems can exist side by side. Wolf uses the example of the movie *Witness* (1985). In this movie, John Book, a policeman on the run played by Harrison Ford, finds refuge in a strictly pacifistic Amish-community (Wolf 1992:792 ff.). When Daniel, one of the Amish, is being harassed and humiliated by an outsider and does not defend himself, Book steps in and hits the aggressor with a fist in the face. Instead of praising Book for his assistance, the Amish Daniel reacts: 'It's is not our way.' Wolf argues that if we agree that both the Amish pacifism and the justice-oriented ethic of the policeman have some plausibility, we adopt another kind of pluralism: pluralism which does embrace relativism without becoming subjectivistic. There may be several moral systems which all have a certain normative appeal and which fall within the range of *acceptable moral codes*, even if their plausibility depends on specific contexts such as religion or culture. The fact that certain systems (like Nazism) fall outside the range of acceptable codes indicates that the criteria for moral rightness are more than subjective [796]. Whereas the first level of pluralism is pluralism within one system and addresses the existence of different values *within* that system, this second pluralism affirms the truth in more than one system. Hence, we may here apply the term, 'systemic pluralism.'[3]

Both kinds of pluralism thus work from the assumption that, despite the fact that opposed moral choices may both be true, their truth is more than subjective only. 'If the subjectivist can be understood as denying the existence of moral truth, the pluralist is better interpreted as believing that, though there may be a moral truth, the truth will be more complicated than one might have wished' [789]. Subjectivism considers moral judgments to be individual expressions of individual attitudes which, apart from being sincere, lack grounds that are valid for persons other than that individual. For the subjectivist, at the end of the day 'anything goes' [786]. The pluralist may find it hard to tell which option is the best but does not say that *all* options are right. What is rejected in Wolf's account of the two pluralisms is not objectivism, but absolutism: the view that in any given situation there is always only one right action or policy.

9.4   Value Pluralism and Systemic Pluralism

It is here that we leave Wolf and, with the help of the distinction between two kinds of pluralism, come closer to exploring their relevance for questions regarding Human ES-cell research.

---

[3] It may be useful to note that the two levels of pluralism are not mutually exclusive: a systemic pluralist can, within her own system, be a value pluralist; the value pluralist can, apart from his own system, accept the plausibility of other systems.

9.4.1   Value Pluralism

Value pluralism implies that in some cases more than one action or policy can be justified on the basis of one and the same set of values.[4] This can have several causes. First and foremost, it is not always possible to realize all values at the same time. Sometimes one value can only be promoted at the expense of another. It may not always be clear what the relative weight is of the values in the case of a collision. In the case of Human ES-cell research we have on one side values such as the search for new treatments for diseases, the development of medical knowledge and skills, the promotion of science and culture, all this backed by the autonomous consent of the donors; on the other side stand the value of embryonic life, the concern to prevent human life from becoming medicalized and instrumentalized, and respect for 'natural processes.'

Whatever the outcome, choices between irreconcilable values often lead to feelings of moral unease. Cancelling an appointment with a friend because something more urgent has come up may cause such feelings, but not cancelling the appointment may cause a similar frustration. Whereas the absolutist suppresses such feelings by reminding himself that he can only have one duty at a time, the pluralist is not so confident. To him, feelings of regret are a reminder of the existence of a plurality of values and of the need to solve future collisions in a way as respectful as possible to *all* values involved.

It is important to note that collisions between values may be settled in more than one way. The first is *prioritizing*: when different values call our attention and we can only promote or save one of them, we have to decide which value will remain unattended. An example may be the decision to save humans first and animals last in a flooding disaster. The second is *threat removal*. This may occur when one value poses a threat to the existence of another, and when it is decided to remove or destroy the threatening cause. An example is abortion of a fetus which threatens the life or health of the mother. A third way to settle a conflict is *sacrificing*. Here, the value that is overridden does not present a threat to the existence of the other value, but the latter would benefit from the purely instrumental use of the former. An example is animal experimentation performed in search of an effective cure for a life threatening human disease. Of these options, sacrificing is the most drastic one and carries the heaviest 'burden of proof.' Solutions in which all values are treated respectfully are clearly to be preferred above solutions which imply the destruction of a value. Since Human ES-cell research involves the sacrificing of a value, we will come back to this below.

Value collisions can take place both within a person and between persons. Evidently, the more persons that are involved and the more complex a community is, the greater is the number of potential collisions. What binds together in value

---

[4] For the sake of brevity, I prefer in this essay to use the term, 'value.' In most cases it can also be read to mean other normative considerations, such as ideals, norms, principles, rights, duties, or virtues.

pluralism is the affirmation of each or most of those commonly held values, even if they cannot all be promoted simultaneously. Different choices may be made, but the parties realise that those who choose differently may do so on the basis of a moral framework which is much like their own.

Value collisions are only one cause for differences in views within a value system. Even when there is consensus which value should prevail, there may still be discussion how this value is to be realized in a complex and changing reality. Dissensus may also exist about the question whether or not new insights from the sciences are allowed to influence our morality. Do findings from modern embryology justify a change in our view on the value of unborn life? Does, for example, the fact that only a minority of natural conceptions leads to a full-term pregnancy imply that embryos have less value than developed fetuses? And does the fact that a fetus cannot experience pain before a certain stage in its development mean that instrumental use before that stage is less problematic? (Cf. Schroten 1988, 2000)

9.4.2 Systemic Pluralism

Much of what has been said about value pluralism can be said about systemic pluralism, but the differences go deeper. Parties take different values to be the 'core' value around which all other values revolve. Values which one party holds and cherishes are irrelevant or even considered wrong by another party. The balances made are consistently and drastically different. We need only to compare arguments based on a Divine command theory (in which 'obedience' is a central virtue) with arguments in which individual autonomy features as the core moral value; or we can compare radical pacifism with just war theories.

Even in systemic pluralism there is some appreciation for those radically different views. For such appreciation to be possible, the different parties must share some of their convictions, especially when it pertains to human rights. But the common basis is thinner. People belonging to different systems are likely to have a harder time to imagine others making different choices because the arguments and motives behind these choices are to a large extent external to them.

Three remarks may be helpful here. First, it comes close at hand to assume that 'value systems' refer to religions, ideologies, or worldviews, but there is no synonymity. Sometimes people who adhere to one and the same worldview seem to belong to different value systems: compare, e.g., the different views which liberal and conservative Christians have on same sex relationships, or compare the different views which Dutch or Swedish Humanists have about assisted suicide. On the other hand, followers of different religions sometimes come remarkably close to sharing one value system.

Secondly, clear cut distinctions between value pluralism and systemic pluralism exist only in theory. In practice, the transition will be fluent. On one occasion people may sense that they share a system of values whereas on another occasion the prevailing sense may be one of alienation. The distinction may nevertheless be useful in order to sound the depth of a dissensus and to explore the chances for reaching some form of consensus.

Thirdly, it is important to notice that we are speaking about plural*ism*, not plurality. That means: the focus is on moral discourses in which the participants have some *appreciation* of the possibility that moral truth may be plural. When the parties insist that only their choice, value, or value system is the right one, we are hardly speaking about pluralism but rather about a plurality of absolutisms. The prospects for a real ethical encounter between such parties are meagre. It should be noted that Wolf herself stresses that her typology does not imply that pluralism in any of those forms is desirable. Unlike her, I would like to make a moral assumption about plurality: discussions about any substantial moral issue with a social or political dimension should preferably take place under the affirmation that morality is pluralistic. When parties exclude the possibility of moral truth outside their own choice, 'moral' disputes will be settled by an 'agreement to disagree,' by a peaceful coexistence, perhaps even by authority, power, or brute force.[5] Pluralism offers a mid-way between the two extremes of absolutism and subjectivism.

Let us thus assume that in order to be able to have a meaningful moral debate about the morality of Human ES-cell research, we are aware that others make different choices and that they have good reasons to do so, that is to say: they are neither epistemically impaired (or simpleminded) nor morally reprehensible (or bad). In a mild form, disagreement as to the morality of Human ES-cell finds its cause in a conflict of values which in themselves are 'good' values; in a more radical form, it is the result of a conflict of value systems which in themselves are 'plausible' systems, i.e., they fall within the 'range of acceptable codes.'

9.5 Moral Empathy versus Moralism

But observing and appreciating the existence of different values and views is not sufficient. Thinking pluralistically needs to be a habit of mind *and* heart. To be able to understand and appreciate the valuing and weighing that others make, we need 'empathy': a capacity to imagine the arguments and motives of the other 'party,' a willingness to discern the truth in them, all in the presupposition of the other party's moral and intellectual integrity.

One of the ways to outline moral empathy is to take a closer look at what I here assume to be its counterpart: 'moralism.' Most definitions of moralism involve notions such as 'narrow-mindedness,' 'conventionalism,' 'knowing what is best for another person,' and 'judgmentalism.' Clearly, it is not enough to define 'moralism' as the habit of telling others what they ought to do. If the friends of an alcoholic try to talk him into seeking professional help, that qualifies them as responsible fellow

---

[5] Of course, such solutions are sometimes unavoidable and helpful, but they should be preceded by attempts to attain mutual understanding of both our differences and our similarities.

humans rather than as moralists. A person's firm conviction about moral truth and moral falseness does not qualify him as a moralist, not even when his truth regards others. Old Testament Prophets such as Moses and Jeremiah are seldom seen as moralists. They are reluctant to announce the Divine judgment over injustice and once they *do* speak, they do so with sadness and tears. Perhaps the moralist can be recognized by the triumphant smile on his face: he rejoices over his own rightness and over the other's wrongness.

This is where our analysis of the concepts of pluralism comes in: the moralist is unable or unwilling to see the values he has overlooked, removed, or sacrificed, but which nevertheless, in a way, are still there. Opinions different from his own, the moralist may find old-aged or modernistic, not to mention stupid or demonic. He is not interested in the arguments and motives of those with whom he disagrees. The moralist may be right in making unequivocal and brave moral choices, but is wrong in assuming that decisions are and should always be supported only by arguments in favor and that an alternative outcome has no positive sides at all.

In contrast to moralism stands moral empathy. Moral empathy neither precludes clear positions nor does it exclude a deep concern that others are making wrong choices. But unlike the moralist, the empathic person knows that morality is a complex field in which choices may sometimes be dilemmatic if not tragic. He knows that his own position may have its own dark sides and cast its own shadows. He knows that not everyone is prepared to make the sacrifices he advocates. Moreover, he is all too aware that he is a fallible human just as anyone else and thus susceptible to making false claims. And he is aware that his opponents are not necessarily at fault – neither morally, nor intellectually.

Although empathy normally functions within social relationships, the moral empathy meant here starts on the individual level. When an individual makes up his mind about a moral issue, this is often the result of a process of discerning and balancing. Even daily actions of minor importance may involve such a weighing. Sometimes it takes a long time to make up one's mind: prior to a decision, one may be torn between opposite directions. When a conviction is reached, this may come as a relief and as a stimulus to focus one's attention on other matters. Still, the values which one has decided to override have not ceased to exist and may thus need our attention and respect. Moral empathy means here: the creative and imaginative effort to remember those values. Its opposite, moralism, means: once having made up one's mind, one reframes the moral issue and rewrites history in a way which affirms the final decision. The moralist thus downplays or forgets the values he decided to override. In hindsight, he wonders about his own narrow-mindedness and rejoices about his newly gained moral insight. Moral empathy is on this level not a social skill; it is a way of dealing with our own values, our own past, i.e., in full respect for the concerns we had before making up our minds.

On the basis of this account of what moralism and moral empathy mean on the individual level, it is possible to imagine what they mean on a *social* and a *historical* level, i.e., when our opinions conflict with the choices of other people in our political and societal community, or when they deviate from choices that were common in times different from our own. Empathy means here: awareness of the values which motivate the choices of others now and formerly (this includes being conscious about the values which steer our own choices). Historically, moral empathy takes the form of awareness of, and respect for one's personal and cultural traditions back in time.

9.6 Moralism as Forgetfulness

Individuals and communities do not always keep in mind the values deemed necessary to override. The Dutch debate about euthanasia[6] before and after its legalization may serve as an example. In 2001, the Netherlands was the first country in the world to adopt some form of legislation on euthanasia.[7] Morally, euthanasia is not much different after 2001 than it was before. In either case, there is a value collision between the patient's well-being and his autonomous wish on the one hand, and moral considerations such as a preference for a 'natural' death, the duty to protect life, and concerns about the professional ethic of the physician, on the other. Despite the ambiguous character of euthanasia before and after its legalization, something close to moralism has occurred in the minds of some.

A few examples may serve to illustrate the point. When the Dutch Parliament voted on the issue, a crowd of 10,000 people demonstrated in protest, an unusually high number for an issue which does not involve socioeconomic interests. The Minister of Health, a fierce advocate of the law, refused to receive a delegation of the protestors. Four days later she declared in an interview: 'Unfortunately, I have lost every form of contact with the opponents, with people who think like they do' (Oostveen 2001). Not much later, a pastor invited me to lecture on the ethics of euthanasia. At my enquiry about his expectations of such a lecture, he confided: 'Well, we all know that no sensible person in this country is opposed to euthanasia, don't we?' I explained to him that the new euthanasia law is not meant to solve the value conflict once and for all: the only thing this law *does* is open just one more option for settling the conflict. On another occasion, the ethics committee of a psychiatric hospital had gathered to develop its policy on physician assisted suicide. When the chairperson proposed a round for inventarizing the policy options for the hospital (from a 'No, never' to a 'Yes,' and everything in between), one member protested: 'Why do we have this conversation? After all, euthanasia is legal now, isn't it?' A fourth example is taken from a social facility in my hometown. When one of the employees, a young man with a mild mental handicap, was told that his boss, who had cancer, had died from euthanasia, he was shocked and reacted emotionally: 'How could this happen? A human person is not a dog!' The deputy chief considered this remark to be so off-limit that he sent the young man on leave for several weeks. Despite the fact that the 'forgetfulness,' or moralism, which typifies all these examples,

---

[6] In what follows, the term, 'euthanasia' means both euthanasia and physician assisted suicide.

[7] The Australian Northern Territories' euthanasia law in 1996 was overruled in 1997. Belgium followed in 2002.

is hardly condoned by any of those who reflect professionally on the topic of euthanasia, the fact that these examples happen, is symptomatic: once a value collision is 'settled' in one way or another, a page is turned and the previous pages are no longer present in the minds of many. History is, even here, written by the winners.

We can only make guesses as to what causes people to forget the values and moral considerations which once played an important role before reaching a decision. Perhaps it is because life goes on and new decisions deserve our attention. It is hard to keep in mind all the considerations we had before reaching a more or less final decision. Moreover, others rightly expect us to stand for the choices we made. To some, showing concern about the values one has sacrificed is tantamount to admitting that one may have been wrong. It runs counter to our sense of pride to express afterthoughts once we have gone through a painful process of making up our minds. Another reason to 'forget' overridden values is of a social nature. Once having made up our minds on a contested policy, we often belong to 'camps' of allies or adversaries. Those who belong to one camp are no longer expected to identify themselves with their opponents. Finally, there are meta-ethical reasons for forgetting overridden values. According to Richard Hare's principle of overriding-ness, there can be no conflict of moral duties since only one duty can be our moral duty.[8] If a maximum of preference satisfaction can only be reached by sacrificing certain values, this is what needs to be done. A similar mechanism may be at work in some versions of a theory of the Divine command: when God asks us to do A rather than B, we need not worry about not doing B.

Still, there are other reasons for *not* forgetting 'where we came from.' The most important one is this: if we agree that values in some way transcend subjective preferences, a value which has become overridden has not ceased to be relevant, let alone ceased to exist. When as a result of a complex weighing process one value or a set of values 'loses,' there is no reason to assume that it should be removed from the moral forum altogether. In that drastic case, moral absolutism would have replaced a free and pluralistic moral debate. Empathy may serve as a heuristic instrument for being aware of the presence of overridden values. Moreover, it can keep us keen on the need to avoid value-clashes in the future or to settle them in such a way that they involve as little value-sacrificing as possible.

9.7 Embryos: Which Values are at Stake?

Many of the values brought forward in the debate about Human ES-cell research are shared by both opponents and advocates. The most important argument in favor of such research is the promotion of the value of health. The importance of this

---

[8] Hare 1984:24. Hare still allows for the possibility that overriding a value leads to feelings of regret. While prioritizing or sacrificing (utilitarians, by the way, do not assume a difference between the two) does not justify a sense of guilt, he argues, we may still feel sorry (Hare 1984:28).

value can hardly be overestimated, especially since a medical breakthrough with the help of Human ES-cells could mean a revolution in the life quality of considerable categories of patients. Moreover, health is a prerequisite for the flourishing of other values, such as happiness and autonomy. Besides, there are the value of scientific progress in knowledge and skills, and the value of cultural progress. Last but not least, there is the economic value of this kind of research: if successful, Human ES-cell research could lead to a 'growth industry' analogous to animal biotechnology and computer technology.

No doubt, the most important value which speaks against Human ES-cell research is related to the human embryo. Here, personhood is both the most important and the most contested value. It is contested because there is no consensus about what we mean by 'personhood,' nor about the criteria for affirming its presence in embryos. And it is important because, if the human embryo is a person, this would justify a ban not only on creating embryos for Human ES-cell research, but perhaps also on any research which destroys embryos. But the concentration on the contested issues of personhood should not keep us from seeing some other considerations. Even if we cannot confirm that an embryo is a person – or if we positively deny its personhood –, we may still have reasons to protect it. Embryos are in some relevant respects similar to human persons; embryos normally stand in one continuous development to human persons; with or without such continuity, embryos are potential persons; they belong to the human species; and they represent the dignity of the parents.[9] Apart from that, we may want to protect human embryos out of respect for natural processes preceding the formation of a human person, out of religious reasons, or out of caution – *in dubio pro embryone* (also known as tutorianism; Damschen and Schönecker 2003:187). We may find it a cultural value to keep certain taboos, just as there is a taboo on eating human corpses. We may also want to protect the embryo out of respect for what people before us have assumed to be the case; or out of respect for those around us whose beliefs we don't share, but whose value system still belongs to the 'range of acceptable codes.' To be sure, when we cannot affirm, or when we positively deny personhood in an embryo, an important reason for its protection is demented. But that does not mean that embryos have ceased to have an intrinsic value at all.

Many of those who advocate the use of embryos for research purposes are still aware of the values they sacrifice (and which they may still be holding). A leading scientist specializing in Human ES-cell research remarked: 'When I look through a microscope and see a human embryo, I experience a hundred times more reverence than if I were looking at an animal embryo.'[10] She displayed the kind of empathic thinking advocated here. Reversely, those against Human ES-cell research should be aware of the plausibility and the urgency of the arguments in favor.

---

[9] Damschen and Schönecker (2003) speak about the so called SCIP-arguments: embryos belong to the human *Species*; there is a *Continuity* between embryos and human persons; embryos are morally *Identical* to grown up people; and embryos are *Potential* people.

[10] The British biophysicist Christine Mummery, in a Dutch radio discussion in 1999.

Although the risk of forgetfulness and moralism thus exists on all sides, we should point to a possible asymmetry here. Values in connection with the embryo may, once overridden, be more at risk to be forgotten than the values of medical and scientific research and the value of health of patients with degenerative diseases. The latter values are not only brought forward by patient organizations and research lobby groups, but are also supported by a broadly shared tendency to define 'morality' exclusively in terms of pain, pleasure, and autonomy. Values connected to embryos lack any of this support. Neither patients nor any group of researchers is likely to benefit from the protection of human embryos.

9.8   Final Remarks

The intensity of discussions regarding Human ES-cell research points to the 'existence' of multiple and sometimes colliding values. There are deep and lasting disagreements as to which values are involved and about the proper way to balance them. Still, all these disagreements do not warrant the conclusion that morality is a matter of mere subjective preferences. As long as the parties agree that some outcomes are wrong, and as long as debates take place on the basis of arguments, there is some awareness that morality has a basis which surpasses the subjectivity of human preferences.

The conviction that the different values brought forward in the debate have an objective basis does not have to lead to absolutism. There may be a 'variety of goods' which collide as a consequence of scarcity, finiteness, and tragedy. Moreover, the human capacity for discerning what is valuable and right is limited and distorted. Thus, debates about Human ES-cell research will benefit much if different parties in the debate keep in mind that rightness is not always an either-or issue, and if they try to empathize with the views held both by others and by themselves in the past. If we try not to forget the values we deemed necessary to sacrifice, we may in the future remain open to the development of alternatives in which both the value of the human embryo and the value of science, medicine and good health are safeguarded.

The objective basis of morality enhances the prospects for a debate on the basis of arguments about a number of questions: what are the presumptions behind the value ranking we make? Are there values we have overseen? What is the effect of sacrificing one value on the respect we have for other values? Can we find scenarios which do not involve the sacrificing of any values? And how do we find policies which are respectful with regard to different views?

Public policies on sensitive ethical issues should in some way reflect this plurality of views and the plurality of values which is present in a society. Let us imagine that a government, after a careful process of deliberation, has chosen a certain policy. Whether this is a 'Yes' or a 'No,' it follows from our analysis of pluralism that such a policy should in some way reflect an awareness of the existence of a value conflict. It should reflect respect for values which were overridden or sacrificed, respect for those who think a different course should have been chosen, respect for the intellectual and moral integrity of one's adversaries, and respect for a culture's own traditions. Thus, a 'no' position should encourage alternatives so as to maximize the health advantage; and a 'yes' position should accept clear limitative criteria and should encourage the development of feasible alternatives which may help to minimize the need to sacrifice embryos. Empathy will prevent overridden values from gliding into oblivion. The debate must go on even after a country, a community, or an individual has made up its mind as to which policy or stance is to be preferred. Empathy and memory keep us awake. They may in the future help us find solutions which do not imply painful and contested sacrifices.

References

Damschen, Gregor and Schönecker, Dieter (eds.) (2003). *Der moralische Status menschlicher Embryonen,* Berlin: Walter de Gruyter.

Gert, Bernard (1988). *Morality: A New Justification of the Moral Rules,* New York: Oxford University Press.

Hare, R.M. (1984). *Moral Thinking: Its Levels, Method, and Point*, Oxford: Oxford University Press.

Hebblethwaite, Brian (1997). *Ethics and Religion in a Pluralistic Age*, Edinburgh: T&T Clark.

Kekes, John (1993). *The Morality of Pluralism*, Princeton, NJ: Princeton University Press.

De Lange, Frits (1995). "Pluralisme en christelijke traditie", *Gereformeerd Theologisch Tijdschrift*, 95.3: 100–125.

De Lange, Frits (2003). "Pal staan voor het pluralisme: pleidooi voor een casco-moraal", *Gereformeerd Theologisch Tijdschrift* 103.1: 40–47.

Oostveen, Margriet (2001). "Minister Els Borst over het tekort van de nieuwe euthanasiewet: 'Ik kan me goed voorstellen dat artsen stervenshulp niet melden'." *NRC Handelsblad*, Dutch National Evening Paper. April 14, 2001.

Schroten, Egbert (1988). "In Statu Nascendi. De beschermwaardigheid van het menselijk embryo vanuit het gezichtspunt van de christelijke ethiek" (Inaugural address), Utrecht: Utrechtse theologische Reeks 4.

Schroten, Egbert (2000). "Waar is dat goed voor? Een theologisch essay over wegen en grenzen in de (bio)technologie." In: Dick G.A. Koelega and Willem B. Drees (eds.). *God & co? Geloven in een technologische cultuur*, Kampen: Kok Publishers, pp. 93–106.

Wolf, Susan (1992). "Two levels of Pluralism", *Ethics* 102: 785–795.

Part III
Philosophical Perspectives

Chapter 10
The Potentiality Argument and Stem Cell Research

Dagfinn Føllesdal

Abstract Stem cell research should be carried out and supported. It promises to yield fundamental new insights in biological processes and has already started to yield some such insights. However, the use of human embryonic stem cells raises serious ethical problems. The ethical status of the embryo is discussed. The view that the strength of our feelings is a measure of the ethical rightness or wrongness of an act is characterized as the most widespread fallacy in bioethical discussions. It also has the unfortunate effect that it tends to replace arguments and thereby stands in the way of a fruitful discussion of the ethical status of the embryo. The potentiality argument is defended against various objections, raised for example in the House of Lords 2002 report on Stem Cell Research. The effort to use adult stem cells and to find ways of deriving young stem cells without destroying embryos is praised.

Keywords Potentiality, argument, feelings, embryo, rights

10.1 Introduction

Biotechnology raises many difficult ethical questions. People differ strongly in their views on what is right and what is wrong. Also the legislation differs from country to country. A German biologist complained some years ago that certain kinds of research on embryos would give him five years in jail, while it might give a British biologist a Nobel Prize.

Many of the ethical challenges in biotechnology are of the same kind as those one finds in many other fields. One must think through a number of questions: What are the consequences of what one does? What are the probabilities and risks?

Stanford University and CSMN, University of Oslo, 0313 Oslo Norway
Present address: Department of Philosophy, Stanford University, Stanford, CA 94305.
e-mail: dagfinn@csli.stanford.edu or: d.k.follesdal@ifikk.uio.no

Who are affected by it? How do they experience what happens? Do they consent to what happens? May they influence the decision? What choices do they have? How are the good and the bad effects distributed between the people who are affected? And so on. These questions are often difficult, but the difficulties usually consist in knowing the facts and the various possibilities. There is largely agreement concerning the normative questions and one must try to find good solutions. The most difficult ethical questions in biotechnology come when we move into areas where we must take a stand concerning the ethical status of human life on the different stages of development, from fertilization to birth.

In this article I will discuss this question and in particular focus on the so-called potentiality argument, perhaps the most used and the most criticized of all the arguments in this field.

10.2   Arguments in Ethics

How can one take a stand on such questions? As a philosopher I am concerned with arguments. In the field of ethics, as in many other fields, arguments are our main way to insight, whether it be concerning questions of right or wrong or questions of true and false.

One first problem we have to face is that many philosophers and scientists in our time have doubted that there can be arguments in ethics. While most of the classical philosophers, from Plato on, held that arguments have a place in ethics, many of the dominant philosophers in the first half of the 20th century maintained that there is no such thing as argument in ethics. This 'no argument' view was found over the whole philosophical spectrum, from the positivists to the existentialists, and it is still held by post-modernists and many non-philosophers, amongst them some scientists and many laymen.

However, amongst philosophers there are very few who now hold a 'no argument' view. This is primarily thanks to the work of John Rawls, Israel Scheffler, Morton White and others, building upon work by Nelson Goodman and W.V. Quine. In particular, Rawls' *A Theory of Justice* (1971) was of great consequence. Here, Rawls introduced the label 'Reflective Equilibrium' for a pattern of reasoning which is common to science and ethics and which he first proposed in an article from 1951 (Rawls 1951). This is the approach that is nowadays used by most people working in ethics, and which I will use in what follows.

10.3   Status of the Embryo

In discussions of the ethics of research on embryonic stem cells it is generally agreed that the key ethical issue is the ethical status of the early embryo. To quote the House of Lords Report on Stem Cell Research from 2002, which I will refer to often in what follows, since it gives the best argued defense for the position which I will criticize

> 4.3 The starting point for consideration of the ethics of research on human embryos is the status of the early embryo. This was one of the most fundamental questions facing the Committee since it is intimately bound up with the questions of when human life begins and of the definition of a person.
>
> *House of Lords 2002*

With this I agree. What then is the ethical status of the human embryo? One view that comes fairly naturally and which appeals to many until one starts to reflect, is the so-called 'gradualist' view: The ethical status of human life at various stages of its development increases gradually from conception or perhaps some later point, until birth. This fits in well with the strength of our feelings. Our feelings for an embryo are weak, and as the fetus begins to grow, our feelings get stronger.

The House of Lords Report favors this view and adduces among other things the following consideration:

> 4.13 A gradualist view of the development of the embryo is also seen as consistent with the way cultures react to early embryo loss. Although would-be parents may feel sad at the natural loss of early embryos before implantation, there is no public mourning ritualassociated with it, nor is there for the loss of surplus embryos left over from IVF treatment.

However, from an argumentative point of view there are serious problems with this view. Already Hume (Hume 1739–40) pointed out that

> The strength of our feelings is no reliable measure of the ethical rightness or wrongness of an act.

To take a contemporary version of Hume's examples: I read a notice in the newspaper that 100 children in Africa have starved to death. I feel sad, but quickly leaf over to the sports page to read the latest news about the Soccer championships. If the paper brings pictures of the children, their fate affects my feelings more strongly, and if I see their death throws live on television my feelings may come in uproar. And of course if one or more of them are my own children, my feelings will be overwhelming. Hume studied the different kinds of relations that strengthen or weaken our feelings in various situations. He thereby used arguments in ethics and differed from many of his followers in the 20th century, whom I mentioned earlier.

Hume's examples show that the hypothesis that strength of feelings is a measure of ethical rightness or wrongness is false. Our ethical intuitions have been shaped through humans living together and interacting through thousands of years. Our intuitions function fairly well when we encounter another human being, especially when we look into the other's eyes and face. It is not an accident that instructors in military close combat warn against eye contact; it can 'demoralize', make it more difficult to kill. And it is easier to kill and bring about maiming, suffering and fear by dropping bombs from a high-flying plane than when one stands face to face with the one who is to be killed. From an ethical point of view I look upon an oppressor's bombing of civilians as a more serious transgression than the suicide bombers of the oppressed. Both activities are morally wrong, but one should not think that because the airplane bomber is less emotionally affected by what he does, his action is therefore less wrong.

Roughly, one may perhaps say that where our feelings are strong and clear, they are a fairly reliable guide to distinguishing right and wrong. Where they are weak and unclear, they are less reliable moral indicators. But both when they are strong and they are weak they are marked by our culture, our upbringing, our inclination toward egoism and a number of other factors which have as an effect that they are not always reliable moral guides.

Our ethical intuitions fail us when we face a test tube. We find it hard to decide what is wrong and what is right. When a human being is killed or injured this evokes strong negative feelings in us, and consequently we regard it as wrong. When a fetus is aborted at an early stage of its development, when it does not look much like a human, our negative feelings are less strong. Accordingly, many look upon this as less wrong. And when we experiment with an embryo in a test tube we hardly have any feelings at all.

Some cells in a test tube, the embryo, and even the fetus do not engage our feelings until we begin getting in contact with them. When the mother notices that the fetus moves, strong bonds start to develop. Also, if one sees pictures of a fetus at various stages of its development, the emergence of eyes and fingers tend to evoke feelings, just as we know that after birth these organs have strong emotional powers. We also know, for example from 'the elephant man' how people who are severely handicapped, so that they do not even look like people, have problems evoking emotional responses. One may say that the fetus, and perhaps also the embryo, are handicapped in a similar way, they do not have the ability to evoke the feelings that are very basic for our protection of human life. And yet, as we shall argue in what follows, they are human life and should be protected.

To show this requires arguments, but such arguments would be irrelevant if one believes that the strength of our feelings may replace arguments. Therefore: *The view that the strength of our feelings is a measure of the ethical rightness or wrongness of an act is in my view the most widespread fallacy in bioethical discussions. It also has the unfortunate effect that it tends to replace arguments and thereby stands in the way of a fruitful discussion of the moral status of the embryo.*

10.4   The Potentiality Argument

Which arguments can we then give for the view that the embryo should be protected? There are several arguments for this, some good, others not so good, but I will concentrate on one, the so-called 'potentiality argument'. This argument goes as follows: *An embryo has the potentiality for becoming a fully developed person; this confers upon it some of the rights of a person, in particular the right not to be destroyed.*

There is much that has to be clarified here, for example, what is meant by potentiality? Which rights are conferred, which not? Some have argued that the notion of potentiality is beyond saving. However, this notion is used daily; it is also used by those who advocate embryonic stem cell research: embryonic stem cells have the potentiality of becoming every kind of cells, while adult stem cells do not have this potentiality. As for the rights conferred, the argument does not say that all rights are conferred, only some rights, which therefore can be considered basic, such as the right not to be destroyed.

A small preliminary remark concerning terminology: The potentiality argument does not say that everything that has the potentiality of becoming a person, is a person. In order to avoid begging the question, let us not call the fetus and the embryo 'persons', for this word is normally applied to beings that are able to think, act and communicate. The embryo is not a person, but it has the potentiality to become a person. We all agree that persons have a right to be protected and the potentiality argument says that that which has the potentiality of becoming a person, should be protected.

A further reason for not using words like 'person', 'murder' and so on when we argue about the ethical status of the embryo is that these words have strong normative overtones; as Locke pointed out, 'person' is a *forensic* term: it has normative implications (Locke 1690, book 2, chapter 27, paragraph 26). Using these words tends to draw our attention away from the arguments: it leads us to quarrel about the use of words rather than focusing on the arguments.

Also the notion of a human life should be avoided when we talk about the embryo. The embryos whose ethical status we are discussing are human embryos. Nobody has contested that. But what does it involve to be human? Thanks to biological research, we know that having a certain kind of DNA is crucial. All that has this kind of DNA is human, whether it be people, or cells of human hair or skin. But we certainly do not call pieces of human hair 'human life'.

In order to have the special ethical standing that concerns us, something more is needed. The philosopher Michael Tooley, in a book *Abortion and Infanticide* (Tooley 1983) has argued that what gives certain collections of human cells their special standing is that they make up an organism with capacity for self-reflection. This may well be so. We all agree that adult human beings with full mental capacities should be treated with respect; we should not take their lives or in other ways hurt them, but rather try to help them along so that they can experience a full and rich human life. However, we intellectuals have a tendency to focus on intellectual abilities. Are there not other abilities that humans have that also contribute to giving us our special moral standing? Kant mentioned the moral law within us as something that filled him with awe, together with the starry heaven above us. It was not our intellectual ability to grasp and reflect upon these things that impressed him, but our situation in this inner and outer universe.

I will not discuss this issue here, but just point out that we would certainly like to include among individuals with a special moral standing humans who lack the capacity for self-reflection, humans who sleep or are in a coma, and we would also like to include children who have – perhaps – not yet developed reflective capacities.

What is remarkable is that Tooley is reluctant to include children in this group. He thinks that the only argument that can be given for including children is the potentiality argument, and he rejects this argument.

To reject the potentiality argument is no easy matter. But Tooley bites the bullet and accepts that it may often be right to kill children, as long as they have not developed the capacity for self-reflection, which, according to Tooley, comes at around two years of age. Of course, we should not kill children haphazardly, but there are situations, for example in the case of children with severe handicaps or with a future with much suffering, where it would be all right to kill them. Generally, the criteria that some people use to justify abortion would for Tooley apply also to very young children. Therefore the word 'infanticide' in the title of his book.

It is a good principle in debates that one should select for criticism the best representatives for the view one criticizes and not choose a representative who obviously is unsatisfactory. Why then Tooley? The reason is that Tooley has seen more clearly than most what consequences one is forced to accept if one rejects the potentiality argument. Many others have taken it for granted that the strength of our feelings is decisive for whether something is right or wrong, and they argue neither for this view nor against competing views. They maintain, for example, that one can accept a kind of physical potentiality, that when the brain and the nervous system have reached a certain level of development, for example, so that an organism can feel pain or think, then it is wrong to take its life. But they rarely argue for their view.

Tooley is clear and consistent. He sees that the potentiality argument or something like it is needed in order to prevent conclusions that few will accept, such as infanticide. And he is willing to take the consequences. He is more consistent than most others who have argued for the morality of ending human life, and he fully acknowledges the costs of giving up the potentiality argument. In my view these costs are so high that they count in favor of the argument. According to the reflective equilibrium approach to ethical arguments, which I use, it counts against a view that it has consequences that go against one's very strong and clear intuitions, in this case the intuition that it is wrong to take children's lives.

What, then, is wrong with the potentiality argument, so wrong that Tooley rejects it at such high costs?

The potentiality argument has been criticized by many, as one should expect, since it is one of the main arguments against terminating the life of a fetus or an embryo. I shall now take up two of the most common objections.

10.5   Potentiality and Other Factors

First, it is often carelessly formulated, especially by its critics. As so often happens in debates, one misstates the view to be criticized. It thereby becomes easy to criticize, but the criticism is directed towards a straw man. The potentiality argument does not state that having a potentiality for a property is sufficient for developing that property. In addition to the potentiality various other factors are needed. Thus, for example, although a newborn baby has the potentiality for becoming an adult human being it is not capable to go through this development all on its own. The baby is dependent on assistance from its surroundings, normally from its parents, or if they fail their duties, then from others who can help it.

This dependence on others is even stronger at earlier stages of development. The House of Lords uses this as an argument against attributing a special moral status to the embryo:

> 4.12. ... Although the fertilised egg and blastocyst contain all the genetic signals required for human life, this is true of nearly all cells in the body. However, genetic elements are not sufficient and there is *no automatic programme of development from blastocyst to birth*. Although the early embryo contains within it the full genetic potential of any person(s) who may develop from it, it requires many other factors, particularly those provided by the maternal environment in the womb, to enable it to realise that potential.

It is not clear what the words I have emphasized mean. Clearly, the blastocyst contains a program for the further development, and if inserted in the uterus, it will follow that program. If the House of Lords by 'program' means something all cells in the body have, then why are they so interested in stem cells? Contrary to what is said in this quotation, some cells have the ability to develop in directions that others do not.

The second half of the quoted passage takes up a different point, with which we all agree: that many other factors are needed, in addition to such a genetic program. Nobody has ever denied this. As in the case of a fetus and a child, the further development of the blastocyst will depend on many other factors. However, as in the case of the fetus and the child, this by itself is not enough to deprive it of a special standing. And it is not enough to exempt those who bring it into being from the responsibility of providing care for it. The responsibility of society is to do its best to see to it that such care is provided, and to assist when assistance is needed.

10.6   Potentiality and Rights

Another objection which is raised against the potentiality argument is that although the embryo, the fetus and the child have the potentiality of becoming fully developed adult persons, this does not confer upon them any rights. This is the most common objection against the potentiality argument, and the House of Lords does not fail to use it:

> 4.10 Those who deny the force of the potentiality argument argue that the fact that a person has the potential to qualify as a member of some class in the future, if certain conditions are met, does not confer the rights that belong to members of that class of being until those conditions are met. A medical student is a potential physician, and if he or she qualifies may practice as such; but the potentiality alone does not confer a right to practice. A child is a potential voter but has no claim to be treated as a voter until reaching the age of 18.

All that these examples show is that many rights, including most of those that are established through social conventions and practices, become effective only when certain conditions are met. The conditions, for example voting age, may vary from country to country. The question is: is the right to *care* and *protection* that human beings are entitled to, a right that comes into effect only at a certain stage in their development, for example birth, or, as in Britain, 24 weeks after conception, or as in some other countries, 12 weeks after conception, or as in the Aristotelian tradition, 40 days after conception for men, 90 days for women. In the legislation of some countries, regulating, for example, inheritance, conception marks the beginning of a legal person. It is mostly taken for granted that the right to care and protection starts when human individual life begins. So we are back to our main question again, when does human individual life begin?

The variety of age limits reflects in part the development in our biological insight. We no longer believe, as did Aristotle, that we start as plants, then turn into animals and finally, in the later stages of pregnancy, become human beings. We now know that we have the same DNA there from conception. In part, the variety of age limits reflects the view that the strength of our feelings is a measure of the moral status of the embryo and fetus. However, as I argued earlier, following Hume, that view is fallacious.

These are the main objections that have been raised against the potentiality argument. More clarification is still needed, and I think that it should be combined with the so-called species argument, that is, the argument that human life in all its stages and all its forms has a special ethical status. I shall not go further into the details of these arguments now, but merely want to conclude this part of my paper by stating that while there is still much work to be done on these arguments, there is hardly anything that resembles arguments on the other side of this issue, for example for the gradualist view. The considerations that are put forth in its favor almost invariably appeal to the strength of our feelings, and as we have seen, strength of feelings is no measure of the rightness or wrongness of our acts.

The burden of proof therefore lies on those who argue for the ethical acceptability of research on the embryo, including embryonic stem cell research.

I will now turn to two other ethical issues connected with stem cell research.

10.7   Priorities, Social Justice

First the emphasis on medical applications of stem cell research. It is often objected against stem cell research that the applications that it is expected to lead to, are very expensive treatments that will benefit only a few patients, who are rich and in the case of many of the illnesses also old. Here is one such objection, from the Conference of European Churches:

> […] we are acutely aware of the global health context and of the plight of readily treatable disease that afflicts so many millions of our brothers and sisters in the poor countries of the world. We draw attention to concerns that expensive stem cell research may be a luxury of 'Northern' lifestyle which expects to live in good health to a good old age. For many of the world's population, living even a full span of life would be a welcome change to their normal expectations. We ask how far stem cell research is justified while European countries promoting it still have not fulfilled their promises of the percent of their GDP they dedicate to aid and support for health care in the developing world.
>
> *To CEC Member Churches, 8 July 2005*

The Chair of the Bioethics Committee of the Israel National Academy of Sciences and Humanities, Professor Michel Revel, wrote about this in another context:

> [...] imperatives of justice and equality in the access to the modern medical technologies such as ES cell research must be upheld.
>
> *Revel 2005, p. 113*

These are nice words, with which we all like to agree. However, given the situation as described by the Conference of European Churches they will remain empty words, at least until they are followed up by concrete proposals concerning how the imperatives of justice and equality are to be achieved.

I am very much in favor of stem cell research. However, I am in favor of it because it promises to yield fundamental new insight in biological processes and already has started to yield some such insight. It is the tradition of researchers to emphasize the practical usefulness of their research, especially when they apply for money. In most countries, Norway included, money for basic research is very hard to get. However, politicians and others are beginning to ask: Where are the cures you promised? Where are the applications? And they will also, like the Conference of European Churches, come up with objections regarding justice and the proper distribution of money for health care, objections that in my opinion are fully justified. In the case of stem cell research, even more than in many other cases, there has been a tendency to oversell the practical applications. Let us emphasize more the theoretical insights and hope that politicians will listen.

10.8   Alternatives

I will now turn to my second concluding issue: Alternatives.

The main argument that is given for embryonic stem cell research is its usefulness, that it opens possibilities for preventing and curing a wide variety of presently incurable degenerative illnesses, such as Alzheimer, Parkinson, and many others. As I just have emphasized, the pure scientific insight that this research is giving is an even more important reason for pursuing this kind of research. Note, however, that they are not arguments for or against the ethical status of the embryo, but considerations that should be weighed against the considerations concerning the ethical status of the embryo that I went through in the main part of this paper.

Given that all the arguments concerning the ethical status of the embryo indicate that it should be protected very much like a born human being, research on embryos is ethically quite problematic. As long as we have no good arguments for attributing a considerably lower status to the embryo, one should try to find alternative ways for obtaining the same insights, alternatives that are not ethically troublesome. In a pluralistic society one should try to find alternatives that all members of the society find acceptable. Where this is not possible, one should at least try to satisfy those views that are well supported by arguments. And the high ethical status of the embryo is at present better supported by arguments than alternative views on the embryo.

It seems therefore to me that one should work hard on developing alternatives to embryonic stem cell research. Research on adult stem cells is one such alternative, against which no ethical objections have been raised. It has been argued that such research has its limitations, and that therefore research on embryonic stem cells should be permitted. However, recently a new alternative has been proposed: *to obtain pluripotent stem cells by a relatively simple process that does not involve the creation and disaggregation of human embryos.*

Important steps in this direction have been taken by Rudolf Jaenisch. In an article in *Nature* he and Alexander Meissner have proposed a way of doing this (Meissner and Jaenisch 2005). Their work has been followed up by others, for example by Deb, Sivaguru, Yong and Roberts in an article in *Science*, 17. February, 2006 (Deb et al. 2006). The idea of this technique, which is called 'altered nuclear transfer,' was originally suggested by Dr. William B. Hurlbut, Stanford, member of The President's Council on Bioethics. (Hurlbut 2004, 2006, see also his paper in this volume).

On Friday, May 5, 2006, two prominent republican members of the Congress, Arlen Specter and the highly conservative Rick Santorum, introduced a bill that would require the National Institutes of Health to research and fund 'methods of deriving young stem cells without destroying embryos.'

These recent developments illustrate two main functions of legislation, with which I will end:

- First, legislation opens possibilities and new alternatives. An example is the restrictive laws on exhaust emission that were introduced some years ago in California. When the laws were proposed, the car manufacturers lobbied against them, arguing that it would be impossible to fulfill them. Within a short time, however, cars satisfying the laws were produced. Something similar seems to be happening in stem cell research.
- Secondly, legislation shapes attitudes. The concern with the status of the embryo and the legislation based on this concern will, I hope, have an effect on society's attitude to all human life from its beginning to its end. If so, the channeling of stem cell research in the direction of alternatives to embryonic research may be a small price to pay for the increased awareness of the dignity of human life.

References

Conference of European Churches, (2005). "New Issues in Stem cells and Regenerative Medicine," p. 4. (Bioethics and Biological Sciences Working Group, Conference of European Churches' Church and Society Commission, 8, rue du Fossé des Treize, FR – 67000) Strasbourg, France. http://www.cec-kek.org/pdf/Stemcells.pdf)

Deb, Kaushik, Mayandi Sivaguru, Hwan Yul Yong, R., and Michael Roberts (2006). "*Cdx2* Gene Expression and Trophectoderm Lineage Specification in Mouse Embryos", *Science 17* February 2006: Vol. 311, No. 5763, pp. 992–996.

House of Lords (2002): *Stem Cell Research – Report*. Session 2001–02. Ordered by the House of Lords to be printed 13 February 2002. http://www.parliament.the-stationery-office.co.uk/pa/ld200102/ldselect/ldstem/83/8301.html

Hume, David [1739–1740] (2000). *A Treatise of Human Nature*. In: Norton, David Fate and Norton, Mary J. (eds.), *Oxford Philosophical Texts*, Oxford: Oxford University Press.

Hurlbut W. B. (2004). "Altered Nuclear Transfer as a Morally Acceptable Means for the Procurement of Human Embryonic Stem Cells", President's Council on Bioethics, Washington, D.C. http://www.bioethics.gov/background/hurlbut.html

Hurlbut, W. B. (2006). "Framing the Future: Embryonic Stem Cells, Ethics and the Emerging Era of Developmental Biology", *Pediatric Research*, Vol. 59, No. 4, Part 2, pp. 4R–11R.

Locke, John [1690] (1975) *An Essay Concerning Human Understanding*. Peter Nidditch (ed.) (Clarendon Edition). Oxford: Oxford University Press.

Meissner, Alexander and Rudolf Jaenisch (2005). "Generation of Nuclear Transfer-Derived Pluripotent ES Cells from Cloned *Cdx2*-Deficient Blastocysts", *Nature* 439 (12 January), pp. 212–221.

Rawls, John (1951). "Outline of a Decision Procedure for Ethics", *Philosophical Review* 60, pp. 177–97.

Rawls, John (1971). *A Theory of Justice*, Cambridge, MA: Harvard University Press.

Revel, Michel (2005). "Ethical Issues of Human Embryo Cloning Technologies for Stem Cell Research." In: Blazer, S. and Zimmer, EZ (eds). *The Embryo: Scientific Discovery and Medical Ethics*, Basel: Karger, pp. 107–119.

Tooley, Michael (1983). *Abortion and Infanticide*, Oxford: Clarendon.

Chapter 11
Can the Distinction between the Moral and the Descriptive Support a Full Moral Standing of an Embryo?*

Øyvind Baune

Abstract One of the issues is to question arguments based upon a clear epistemic split between the normative and the descriptive meant to support the full moral standing of embryos, especially that the concept 'human embryo' is a purely descriptive concept independent of moral considerations. Two arguments in support of this view is criticized, i.e. the metaphysical and the potentiality argument. The present view is not only that an alleged descriptive *premise* may be co-conditioned by normative considerations, but that even *concepts* used in such premises, e.g. what count as human embryos, are co-conditioned by moral considerations. Concepts of this type correspond (roughly) to what Bernard Williams called 'thick' concepts. It is argued that these concepts express natural properties, that they can figure in the norm conditions of general *prima facie* ethical norms, and typically express gradual properties. The consequence of this is a gradual view according to which the strength of the ethical considerations to protect a human embryo does vary and increases to a full moral standing some time during pregnancy. – At the end a further argument against a non-gradual moral development is added.

Keywords Graduality, human being, potentiality, reflective equilibrium, thick concept

11.1 Introduction

The use of human embryos has a great potential, both in medical research and in medical treatment. If such uses are unethical, convincing arguments are hence needed to show this. What seems to be the only possible arguments in support of

The Department of Philosophy, Classics, History of Art and Ideas, University of Oslo.
e-mail: oyvind.baune@IFIKK.uio.no

* Thanks to the participants of the project 'The moral status of human embryos with special regard to stem cell research and therapy' for very helpful comments upon earlier versions of this paper.

this are those that can show that the moral dignity (or moral status) of the human embryo is incompatible with such uses. Let us take for granted the moral status of grown up human being. If there is no ethical relevant difference between a human embryo and a grown up human being that does support a different moral standing between these, the embryo has the same full moral standing as a grown up human being. If this is the case, it would be unethical to use an embryo 'only as a means' (to use Kant's terminology) – as it actually would if used in research – however promising this would be from a medical point of view.

Some of the arguments supporting a high moral status of even an early embryo are based upon a clear split between the descriptive and the normative. An aim of this paper is both to have a closer look at this and to see what alternatives there is to this. Given this difference, what kinds of ethical arguments are these that support the moral status of an embryo? Some of those who reject any use of embryonal stem cells do this on a Christian basis. I shall, however, start the discussion with what I take to be a traditional an argument, which due to its formality, can be interpreted both as a Christian and as a philosophical argument in support of the full moral standing of the embryo from conception.

11.2   A Traditional View of the Normative and the Descriptive

The first argument I shall consider here, which in the outset seems reasonable, is an argument based upon two premises both of which with an alleged strong intuitive appeal, one *normative* which says:

P$_1$: A human being has a full moral status,

and the other *descriptive*, which says:

P$_2$: A human being starts at conception,

from which (logically) follows the normative conclusion:

C: A human being has a full moral status from conception,

and hence that the killing of an embryo is morally wrong.

There is, however, an important underlying presupposition in this argument, namely that there is a clear epistemic difference between the premises in the sense that the truth of the *descriptive* premise P$_2$ is a purely scientific concern, i.e. supported (or rejected) on a purely descriptive scientific basis, while the *normative* premise P$_2$ is immune towards scientific arguments and hence supported (or rejected) on a purely normative, non-scientific basis.

The first aim of this contribution is to show that this type of argument – which I take to correspond to the traditional understanding of the distinction between the normative and the descriptive (at least after Hume's is/ought distinction) – does not hold good.

But before doing this, one should note that because this traditional view presupposes a clear distinction between the normative and the descriptive, and a corresponding mutual epistemic independence of these types of premises, this type of argumentation can be based on various types of normative premises, whether theological or philosophical. The critical comments below against this view is general and meant to be independent of what type of – or epistemic basis for – the normative premise in such argumentation.

A crucial question concerning the 'P$_1$ & P$_2$, hence C'-argument is how to understand the concept 'human being' as it figures in P$_1$ and in P$_2$. Logical validity of the argument presupposes that it is the same concept 'human being' – or more precisely: concept(s) with the same *extension* – in both premises. Otherwise, it is a logical flaw. But which extension is this?

That a human being has a full moral standing may seem obvious, and even intuitively correct, at least from birth. Hence, premise P$_1$. Next, that a human being is *the same entity* from its conception, through pregnancy, and after birth, may seem reasonable.[1] So, since this entity, at least after birth, is a human being (P$_1$), it may seem reasonable – due to being the same entity from conception to after birth – that it is a human being already from conception. Hence, premise P$_2$, which, together with P$_1$, implies C.

This I take to be the core of the argument on which I now shall make a closer examination.

11.3 Critical Comments on the Traditional View on the Normative and Descriptive

A problem with this is how to come to terms with the extension of the common concept 'human being'. Let us start with premise P$_1$. Since P$_1$ is supposed to be independent of P$_2$, this means that P$_1$ presupposes that this extension does not include anything which does not have a full moral status, which means that it is a *moral* question which extension it is reasonable to ascribe to the concept 'human being' in P$_1$ – at least that morality puts a limit on this extension. The truth of this premise presupposes that the extension of 'human being' is within the extension of the concept of having 'a full moral status'. But since the concept 'human being' in P$_2$ is supposed to have this same extension, also P$_2$ is based upon this implicit moral presupposition, which hence makes also P$_2$ dependent upon a moral consideration. So, P$_2$ can *not* be understood as a purely descriptive premise, but rather as having also an epistemic *normative* basis – at least if the argument is supposed to be logically valid.

To express this simpler one could say that the extension of the concept 'human being' in P$_1$, because P$_1$ is an ethical premise, is co-determined by ethical considerations. And since logical validity presupposes the same concept 'human being' in P$_1$ and P$_2$, also the extension of 'human being' in P$_2$ is co-determined by ethical considerations, which is incompatible with P$_2$ being a purely descriptive premise.

---

[1] Exceptions from this are identical twins and re-union of twins.

Now, since this did not work, let as next see if the argument fares better if we start with the premise P_2. Since P_2 is supposed to be independent of P_1, this means that it is a purely descriptive (scientific) question where to draw the borderline of the extension of the concept 'human being'. But then the problem is that P_1 just claims, or presupposes, without any argument, that every entity within the P_2-extension has a full moral standing. This is hence in need of an argument – especially since this seems counter-intuitive – which is neither given with P_1 nor with P_2.

A third possibility is to take for granted the intuitive appeal that P_1 and P_2 have independent of each other. But then the question remains whether the extensions of 'human being' each of which makes the premise P_1 or P_2 intuitively acceptable, are identical or not. If this identity is what the argument presupposes, it makes it circular, and if this is not presupposed, it is in need of an independent argument, which is not given.

The upshot is that however one tries to come to terms with this argument – i.e. whether based upon the intuitive appeal of P_1 (which makes P_2 problematic), or upon the intuitive appeal of P_2 (which makes P_1 problematic), or based upon the intuitive appeal of both P_1 and P_2 (which makes the argument circular or logically invalid) – it comes out badly.

Even if this does not show that the argument necessarily is logically invalid, or that the premises are untrue, it undermines the 'strong intuitive appeal' of the premises (as the argument presupposed). This means that the premises P_1 and P_2 are in need of an independent justification. In what follows two attempts to do that will be scrutinized.

11.4 A Problem with the Metaphysical Argument: Created in the Image of God

Let us in the followings, for the argument's sake, grant both the truth of P_2 and that it is a purely descriptive premise. This presupposes that the extension of the concept 'human being' includes every human embryo from conception and onwards. Then the question remains: how to justify the normative premise P_1 given this understanding of 'human being'? One possibility is to base it on a metaphysical premise, e.g. that a human beings is *created in the image of God* or is being *ensouled* (*animated*).[2] A metaphysical property in this sense can be defined as a property the existence of which does not make any empirical difference, either directly or indirectly (i.e. in terms of the hypothetico-deductive method). This means that one can neither argue for, nor against, such a property on a purely scientific basis. This does not, however, rule out the possibility of basing one's argument concerning the moral status of human beings on a metaphysical premise, given that this is supported in a different, non-scientific way, e.g. by the Christian revelation.

---

[2] The latter, i.e. ensoulment at the time of 'quickening' (i.e. when the fetus starts moving), is a traditional Catholic view. The following discussion of this argument is independent of just which of these (or some other) metaphysical concepts is used.

Even if granted that such a metaphysical property (e.g. *created in the image of God*) is sufficient for ascribing a full moral status to entities with this property, it raises the following problem: A metaphysical property (in the sense suggested) can not in itself be (or yield) an applicable criterion of deciding who shares this property. In order that a metaphysical property can be practicable one needs a criterion which connects the metaphysical property with some empirically applicable rule. A possibility could be: 'A being *created in the image of God* starts *at conception*' or: 'A human embryo gets *ensouled* when it *for the first time makes a movement.*' This connects the metaphysical property: *created in the image of God*, or: being *ensouled*, with an empirically applicable criterion: *at conception*, or: when the mother *for the first time feels a movement in her womb*. Without a rule connecting the metaphysical non-observable property with some observable criterion, the metaphysical criterion cannot in practice work.

But this raises the following problems: which rule is the correct one, i.e. at what *time* (at conception? or at some later time during pregnancy?) does a human embryo obtain the metaphysical property at stake and hence get a full moral standing? and: how can this point of time be justified? Such a criterion can not be empirically justified and must hence be based upon a metaphysical justification. There seems, hence, to be no alternative than based upon a supernatural revelation. But what kind of revelation could that be? It can not be the Bible since that the identity of the human fetus goes back to conception was first known in 1827, and was hence unknown during (most of) the church history. So, how can this metaphysical point of view then be given a theological justification? I do not see how.

There is a further problem with this metaphysical view which I will come back to at the end.

11.5   Which Type of Potentiality?

There is, however, another argument which (like the previous ones) presupposes a clear split between the normative and the descriptive, but (unlike the previous ones) is based on a non-metaphysical premise. That is the so-called potentiality argument. Like the previous one it grants the truth of P_2, and gives an argument in support of P_1. On this occasion I shall only give a sketch of it and (try to) show that it raises a variety of alternatives with no clear or unique understanding of potentiality.

A main version of the potentiality thesis says that there is no difference in intrinsic value between an embryo from conception, through pregnancy, until being a grown up human being. The alleged reason for this is that there is no morally relevant difference between being a grown up human being in *potentiality* and in *actuality*. The embryo, qua a potential person, hence has the same moral standing, from conception and onwards, until being a person in the full sense.

Some has argued against this by giving examples of entities with different values depending upon having a certain property in potentiality and in actuality, e.g. being an acorn compared to being an oak tree (Singer 1979:120; Williams 1995:221).

But neither this nor similar examples can in a reasonable sense be generalized (as Singer seems to believe) to a thesis that there *in general* is a difference in value between potentiality and actuality. A problem with this is that it overlooks that there are different types of potentiality associated with different types of values. The difference in values between an acorn and an oak tree – which is both Singer's and Williams' example – is in *instrumental* values. But there may also be a difference in instrumental value between humans, say, between a physician and a philosopher (like me) during an accident when people are seriously hurt. But a difference in instrumental value says nothing about whether there is a difference in *intrinsic* value between these humans.

But even if the question concerns the same type of *value*, e.g. the intrinsic value (generally ascribed to humans), there still are different *types* of potentialities. Having a conception of oneself, a moral understanding, a rational attitude towards the future, etc. is generally considered a sufficient basis for having the full moral standing as a human. But is it a sufficient condition for this standing to have these properties only in potentiality? That seems to vary. That this potentiality is sufficient for the full moral status of an unconscious patient during an operation, is unquestionable (Byrne 1988:94). But that is a different type of potentiality than the 'potentiality for normal development into a human being', as Williams put it (Williams 1995:221), which is the present concern. Does this development-into-a-human-being-type make a morally relevant difference between potentiality and actuality? There are reasonable examples that even this 'development' type of potentiality ascribes a full moral standing to the entity in question, e.g. a human baby: Even though the actual level of mentality of a calf (to use Singer's example[3]) is higher than the actual level of mentality of a newborn human baby, it is lower than the potential level of mentality of the baby. If a baby has a higher moral standing than a calf – which (with a few exceptions[4]) seems to be generally accepted – this presupposes that potentiality does count.

There seems, however, to be several alternatives to come to terms with this. One is that potentiality, concerning a full moral standing, starts at the conception, another that this moral standing goes back to the 'primitive streak' or to the beginning of the cell differentiations,[5] a third that it is a gradual process in which the embryo reaches its full moral standing some time during pregnancy. The potentiality theses – even if restricted both to the 'development'-type and to

---

[3] Cf. 'For on fair comparison of morally relevant characteristics [...] the calf, the pig and the much derided chicken come out well ahead of the fetus at any stage of pregnancy' (Singer 1979:118).

[4] Exceptions include Peter Singer (cf. previous note) and Michael Tooley.

[5] Even if accepting the potentiality argument, Byrne has adduced an argument to undermine that potentiality goes all the way back to conception: 'the embryonic matter in the days immediately following conception [...] is not even differentiated into that which will become the fetus and that which will develop into the placenta', Ibid., p. 99. – Farley claims that a '[g]rowing number of Catholic moral theologians [...] do not consider the human embryo in its earliest stages [...] to constitute an individualized human entity with a settled inherent potential to become a human being' (Farley 2001:115).

intrinsic value (cf. above) – seems still to be compatible with several alternatives and hence does not in itself give a unique and clear answer to which of these is the ethically preferable one.

The only way to solve this problem seems to be to use the method of reflective equilibrium. This will be a topic later in the paper. Before this, an argument which takes a step in the direction of 'combining' the normative and the descriptive will be examined.

11.6   Morton White on the Normative and the Descriptive

The 'P$_1$ & P$_2$, hence C'-argument was based on a clear epistemic split between the normative P$_1$-premise and the descriptive P$_2$-argument, i.e. that the reasons for accepting P$_1$ and for accepting P$_2$ were independent of each other. The following argument comes close to this in accepting the distinction between a normative and a descriptive premise, but deviates from this in accepting a mutual epistemic dependence between the premises, especially that a normative consideration may support the descriptive. An argument which comes close to the 'P$_1$ & P$_2$, hence C' - argument is an argument given by Morton White. The following is a slightly simplified version of his argument (White 1981:30):

1. Whoever takes the life of a human being does something that ought not to be done.
2. Every living fetus in the womb of a woman is a human being.[6]

Hence, given the normative premise (1), which can be considered to be a consequence of P$_1$, and the descriptive premise (2), which can be a consequence of P$_2$, it follows that:

3. Whoever takes the life of a living fetus in the womb of a woman does something that ought not to be done,

which also can be considered a consequence of C.

According to White, (1) is a normative premise, (2) is a descriptive premise, while (3) is a normative conclusion. His reason for something being descriptive is 'because it contains no «ought» or «may»' (Ibid.), which (2) does not.

But the truth of the conclusion is not established with this argument even if the argument is logically impeccable. What is shown is only that *if* the premises are true, so is the conclusion, and: *if* the conclusion is unacceptable (at least) one of the premises has to go. But if so, which one? White's point – which is an extension of an understanding of scientific falsifications going back to Duhem and Quine – is

---

[6] White's original premise was the following: 'Every living fetus in the womb of a human being is a human being' (Ibid.). I changed this to make the logical form of the argument clear such that 'human being' is the common concept in (1) and (2) that does not appear in (3).

that *if* the conclusion of the argument is rejected, it only follows that (at least) one of the premises has to yield, but that the argument does not tell *which* of the premises this is. What in this which goes beyond Duhem and Quine and beyond the logic of scientific falsifications, is that the conclusion is normative, while the set of premises (one of which may be rejected) includes both a normative and a descriptive premise. If this White's generalization of Duhem and Quine's 'holistic' view is acceptable (which I think it is) it follows that if the normative conclusion is rejected, it is logically permissible to reject any of the premises from which this conclusion follows, including the descriptive one. The rejection of the normative conclusion can hence be used to 'falsify' a descriptive premise. This is, hence, to let a normative thesis (= the negation of the normative conclusion) support a descriptive thesis (= the negation of a descriptive premise), and more generally, to accept that the truths of descriptive theses can be co-determined by the truths of normative theses.

So far this seems acceptable. But there is still a problem with this, viz. whether 'human being' – which logically connects the alleged normative premise (1) and the alleged descriptive premise (2) – is a purely descriptive concept or has in addition a normative aspect tied to it which co-determines its extension. That (2) is a purely descriptive premise – as White apparently thought – presupposes that 'human being' is a purely descriptive concept in the sense that no ethical consideration affects, or is needed to decide, whether an entity is a human being or not. The truth (or falsity) of (2) is hence only dependent upon the extensions of the concepts 'living fetus in the womb of a woman' and 'human being', both of which have to be purely descriptive in order to satisfy White's condition that (2) is a descriptive premise.

But this raises the following problem: How can the descriptive premise (2) be rejected on a normative basis – due to the rejection of (3) – when the truth of (2) is completely determined by its descriptive concepts because (as White expressed it) 'it contains no «ought» or «may»'? The rejection of (2) presupposes *either* that the extension of 'living fetus in the womb of a woman' is too large *or* (rather) that the extension of 'human being' is too small to satisfy the truth-condition of (2). But then the rejection of (2) means the rejection of (or a modification of) at least one of these extensions – which is *due to a normative consideration*: Whether its extension is changed or its original version maintained is in any case then due to normative considerations. This is hardly compatible with considering 'human being' as a purely descriptive concept. And if not, neither is the premise P_2 a purely descriptive premise. I fail to see how White can come to terms with this.

One should note that this has a similar consequence also for the concept 'human being' in the 'P_1 & P_2, hence C'-argument, and, hence, for this very argument. Also other concepts, like 'person', raise a similar problem. The question is whether there is *any* acceptable concept – which logically connect the normative and the descriptive premises (1) and (2) – which do avoid this problem. That would presuppose that its extension is independent of any normative considerations even though this concept is part of both a normative and a descriptive premise. But how can a rejection of such a normative premise with this alleged descriptive concept avoid having an effect upon the extension of this concept?

The upshot of this is *either* to give up White's idea of 'falsifying' a descriptive premise due the rejection of the normative consequence *or* to go one step further than White and accept that normative considerations may have an impact upon the concepts of the descriptive premise, which points towards 'thick' concepts (see below). Which of these to opt for, will be the outcome of using the method of reflective equilibrium. This we will come back to after a brief introduction to thick concepts.

11.7   Thick Concepts

What seems to be an even better understanding of the normative/descriptive distinction is given by Bernard Williams (even though he prefers the more appropriate distinction between the *ethical* and the *scientific* (Williams 1985:135)). A crucial point is what he called 'thick' concepts. Examples of this are 'such as *treachery* and *promise* and *brutality* and *courage*, which seem to express a union of fact and value.' (Ibid.:129). He later added '*coward*, *lie*, *brutality*, *gratitude*, and so forth' (Ibid.:140).

The alternative traditional view about such a concept is that it has both a descriptive and a normative sense, but that it is its descriptive content which determines its extension which hence is independent of its normative content. This means that which actions are, say, *cowardly*, *lying*, *brutal*, etc. can be decided without sharing, and even understanding or knowing, the evaluative contents of such concepts. People using such terms can hence agree *what* are treacheries, promises, brutalities, etc., but still completely disagree about the *evaluations* of such acts.

A criticism of this view – Williams was not the first to express this – is that the determination of the extensions of these concepts is not merely a descriptive task, but co-determined by evaluative concerns. Without these ethical aspects, the extensions of these concepts can not be decided.

To clarify how to apply a thick concept, let us first note that the descriptive meaning of a thick concept, or rather its extension, is not one-sidedly determined by its evaluative meaning. Thick concepts have extensions co-determined by both their descriptive and their evaluative meaning.

Secondly, it seems reasonable that a thick concept not necessarily is a sortal (classificatory) concept, but that it may express – perhaps even typically express – a property which comes in degrees,[7] e.g. being *brutal*, *coward*, or *kind*.[8] These degrees are presumably conditioned by both the *type* of the underlying subvenient property and its *degree*.

---

[7] Cf. Quine's distinction between 'mass terms' and terms with 'divided reference'. The first type can be counted, the other measured (Quine 1960:§ 19).

[8] All of the examples previously quoted from Williams I take to come in degrees in the sense that for each of them its degree of moral badness or goodness may vary.

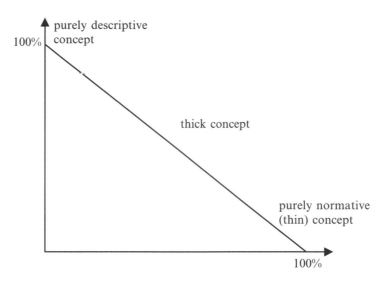

Fig. 11.1

Thirdly, I take it that a thick concept are located somewhere in between a purely descriptive concept, e.g. being a *stone*, and a purely evaluative (thin) concept with no descriptive meaning, e.g. being *good* (cf. Fig. 11.1[9]). That means that the mutual amounts of evaluative and descriptive contents can vary.

Now, let us apply this to the 'P$_1$ & P$_2$, hence C' – and '(1) & (2), hence (3)'-argument. Is 'human being' a thick concept? The previous discussions of these arguments do support this:[10] Since neither of the premises turned out as either a purely normative or a purely descriptive premise, this supports that the concept 'human being' actually is a thick concept. (A proposition cannot be normative without including a normative concept.) This, then, can explain why neither a purely descriptive nor a purely normative understanding of either of these premises does work.

Secondly, is 'human being' a classificatory, either-or property or a property that comes in degrees, say, which increases during pregnancy? Since the answer to this,

---

[9] This graph is meant to express the 'sum' of the descriptive and the normative contents of concepts, whether being *purely descriptive* (top left), *purely normative* (evaluative) (bottom right) or being *thick* (somewhere in between), given that it is possible to quantify the respective percentages of the descriptive and the evaluative contents of the concepts. Whether this is possible, I do not know. But if it is, the graph will be a straight line as shown in the figure.

[10] I have, though, not found 'human being' in Williams' lists of thick concept. On an occasion he argues that a human being does not start at conception without mentioning whether this is a thick concept (Williams 1995:221 f.).

which is crucial concerning the ethical status of embryos and fetuses (if granted that those are thick concepts), is dependent upon a precise understanding of ethical method, it will be postponed until after that method has been more thoroughly clarified.

Thirdly, clearly 'human being' is neither close to being a thin concept nor without a descriptive content, but located in between the purely descriptive and the purely normative locations (cf. Fig. 11.1).

So far, this was about the concept 'human being'. But similar considerations are relevant for a lot of other concepts, like 'person', 'human subject', etc. A consequence of this is that if one is discussing the moral standing of such entities, it seems impossible to avoid taking the thick aspect into consideration. Our moral intuitions are strongly related to such thick entities. A crucial question is whether thick concepts can be substituted with neutral concepts as they figure in the norm condition of general ethical norms. If one accepts that moral intuitions do count it seems difficult to refuse to take thick concepts into account.

11.8 Thick Concepts and the Method of Reflective Equilibrium

The idea of supervenience, i.e. that evaluative concepts (or properties) are conditioned upon descriptive concepts (or properties) such that there cannot be a normative difference without there being a corresponding descriptive difference which accounts for the normative difference, seems intuitively correct. A controversial issue, though, is how far the supervenience thesis goes. If this is understood in accordance with Hare's prescriptivism the extensions of concepts like *brutal, coward, kind*, or the like, are one-sidedly based only upon the descriptive meaning of these concept and independent of their evaluative meaning. But this is hardly compatible with thick concepts, since their extensions are co-determined by their evaluative senses.

Does this mean that we have to give up the idea of supervenience? In Hare's sense, yes. But there is an alternative way to come to terms with this, i.e. how to make supervenience compatible with thick concepts. A crucial point is how to understand that the normative is 'conditioned' upon the descriptive. This can mean two different things, or rather that there are two aspects of this. To come to grips with this, it can be compared to logic: Analogous to that a set of premises – according to 'top-down' method – can justify a logical conclusion, similarly a set of descriptive properties can condition a normative property. (So far, this is in agreement with Hare, but not the following:) And analogous to that a logical conclusion – according to the 'bottom-up' method – can inductively justify (or at least contribute to the justification of) its premises, a normative property can co-determine its descriptive property. The latter means that the normative meaning of the concept has a role when deciding its extension.

But note that this says nothing about what is the *epistemic* primary premise, whether the descriptive or the normative (analogous to that a logical argument in itself says nothing about whether it is the premises or the conclusion which is the epistemic primary subject). I think this is the core of the problem with Hare's account: Even if, as Hare reasonably enough presupposes, that the normative is conditioned by the descriptive this does not rule out that one, in order to come to an agreement between the normative and the descriptive sense of a thick concept can approach this *either* from a 'top-down' *or* from a 'bottom-up' perspective. This latter approach means that the descriptive meaning of the concept is modified in the light of its possible normative meaning, a point which is compatible with the idea that the normative is conditioned by the descriptive.

This, so far, is a contrast between a 'top-down' and a 'bottom-up' understanding of the descriptive and the normative. But an even better approach is a combined approach *both* from a 'top-down' *and* a 'bottom-up' perspective in agreement with the method of reflective equilibrium. This means that it is a mutual epistemic dependence, or conditioning, between the evaluative and the descriptive aspects of such concepts.

The upshot of this is that in analyzing thick concepts, in particular how to come to a reasonable understanding of their extensions – which is the crucial question in the discussion of the moral status of human beings – both its normative and descriptive senses have their say. Thus, if a suggested extension of a thick concept clearly clashes with one's normative understanding of it, one of them has to yield, either the extension being changed or the evaluative meaning being modified or even given up.

11.9   Can Thick Concepts Be Avoided?

The question now is whether one can avoid thick concepts and base one's reasoning – especially the norm conditions of one's principles – on concepts with extensions independent of evaluative considerations. Given the previous argument that a mutual epistemic dependence between the normative and the descriptive premises presupposes thick concepts, this avoidance implies the rejection of any mutual epistemic dependence – even in White's sense – between the normative and the descriptive premises. A consequence of this is that concepts which figure in both the normative and the descriptive premises do have extensions independent of moral considerations.

What I take to be the problem with this is that it takes for granted that the process towards a reflective equilibrium has no effects upon the extensions of the concepts involved, especially the common concept in the normative and descriptive premises, e.g. 'human being' in White's argument. But how can the normative premise be modified or replaced (due to a process towards a reflective equilibrium) without either changing or modifying the extension of the common concept in the argument? In both cases the extension of the resulting concept has been changed – due to moral considerations. The outcome is a concept (and norm-condition) whose extension is co-determined by moral considerations – as thick concepts are.

11.10   Prima Facie or Everything-Considered General Principles?

This essay is based upon the method of reflective equilibrium. But as already Rawls was aware, there are 'several interpretations of reflective equilibrium' (Rawls 1971:49). I shall here argue for a version of this which deviates from Rawls' version of this method, but is closer to, but not identical to, the variant favored by Beauchamp and Childress. This corresponds to a version of what Rawls called 'intuitionism' which is a theory based upon several principles which in particular cases may yield conflicting suggestions but without a priority rule for weighing the conflicting considerations against each other. I shall on this occasion restrict my comments to just one point. A vital point is how to understand the very conception of general *prima facie* ethical principles and how they are applied to a variety of cases.

As mentioned, intuitionism accepts that it is not possible to obtain an ethical theory which in a strict, non-intuitionistic sense, yields conclusions which without exceptions matches our considered judgments. But this does not rule out that one *to some extent* can obtain a theory which in a logically strict sense can be applied to various cases. The present view is a theory based upon a set of general *prima facie* ethical principles each of which ascribes a certain *prima facie* ethical property – say, *prima facie* ethically permissible – to every case which subsumes under it. An example is 'Every killing of a human being is *prima facie* wrong'. One and the same case may subsume under different *prima facie* principles, each of which ascribes different *prima facie* ethical properties to the case, say, being *prima facie* forbidden and *prima facie* obligatory. The final ethical property, i.e. the everything-considered ethical property, is then reached by balancing these *prima facie* norms against each other. This presupposes the use of judgment, which represents the non-strict aspect of this method.

To see this, let us take the example of the killing of Martin Luther King, which obviously was a morally wrong action. But why? Was it because it was a killing of a *human being*? But there are killings of human beings that are morally acceptable, say, if done in self defense. But what about the killing of an *innocent* human being, as the killing of Martin Luther King obviously was? But Bernard Williams has given an example that even the killing of an innocent human being can be morally acceptable (Williams 1973:96 f.).

A consequence of this is that since the killing of Martin Luther King was immoral, the relevant aspects of this action include the conjunction of the negations of every set of properties each of which would be a sufficient moral reason for accepting the killing of a human being. And since the list of these negations presumably is open-ended, it presumably is impossible to obtain the totality of ethically relevant features of the killing of Martin Luther King.[11] And if it, contrary to this, still would be possible to obtain this, this list obviously would be long and complicated. The upshot is that starting with the Martin Luther King example, whose ethical property is obvious and non-controversial, the attempt to obtain a complete list of the relevant properties of this case is very difficult, probably impossible. And since the availability of this list is a presupposition for expressing the general ethical norm under which the killing of King subsumes, it follows that even this presumably is impossible. This means giving up the prospect of obtaining general ethical norms which without exceptions ascribe the final – i.e. everything-considered – ethical properties to the cases under which they subsume. The alternative is to rely upon general *prima facie* principles.

11.11   Prima Facie Principles and Natural Properties[12]

But that this argument counts against the attempt to express – and then rely upon – general everything-considered ethical norms, does not mean that it is not possible to take into account all relevant aspects of a case, like those of the killing of Martin Luther King. But this is done simply by subsuming the case under general *prima facie* ethical principles. The present point is that the relevant aspects sufficient to decide the case at issue can be accounted for by subsuming it under one or a few general *prima facie* norms.

How can this be the case when the totality of relevant aspects is so complicated and difficult to come to grips with? A crucial point in this is the concept of *natural* property. The idea is that the relevant features which are natural properties are sufficient to decide the case. A consequence is that the norm conditions of general *prima facie* principles, in contrast to general everything-considered norms, do include only natural properties – or conjunctions of natural properties – not the negations of natural properties (which typically are not natural properties). Presumably, the only relevant natural property of the killing of Martin Luther King is the *killing of a human being*. All further relevant aspects of this action are *negations*

---

[11] The underlying logic of this is a generalization of J. L. Mackie's concept of INUS-conditions (originally meant for causal conditioning) to be applicable to the norm condition of general ethical norms: Let $R_1\&R_2$ and $R_3\&R_4\&R_5$ and..., be the conjunctions of *all* sets of relevant conditions each of which (e.g. $R_1\&R_2$) is a sufficient condition for making an exception to the prohibition of the killing of a human being. Then $(R_1\&R_2\lor R_3\&R_4\&R_5)\lor...$ is the total condition making the killing of a human being permissible. The negation of this, i.e. $\neg(R_1\&R_2)\&\neg(R_3\&R_4\&R_5)\&...$ or $(\neg R_1\lor\neg R_2)\&(\neg R_3\lor\neg R_4\lor\neg R_5)\&...$, is then the total condition for *not* making the killing of the human being morally permissible. This obviously is a very complicated condition of which it hardly is possible to give a full explication.

[12] The present point is a generalization of Peter Gärdenfors' theory of inductions (a generalization which he seems to have foreseen), namely that inductive generalizations is based upon natural properties (Gärdenfors 1990:87 ff.).

of natural properties, e.g. that the killing was of an *innocent* human being, and hence are not natural properties and does not figure in any general *prima facie* norm. The only general *prima facie* principle under which this killing subsumes is hence the one saying that it is *prima facie* morally wrong to kill a human being. But since the further relevant aspects are negations of natural properties, these natural properties figure in *other* general *prima facie* principles under which the killing of Martin Luther King does *not* subsume and hence has no say concerning the everything-considered ethical properties of the killing of Martin Luther King.

The idea is that general *prima facie* principles under which a case subsumes is a sufficient basis for the use of judgment to reach the everything-considered ethical property of the case, and that the norm conditions of these are natural properties which typically are thick concepts.

11.12   Graduality of Ethical Considerations

A final point is this. Does it follow from this that every case that subsumes under the same set of general *prima facie* ethical norms, ends with the same everything-considered ethical property? If so, it would be possible to express this in general, everything-considered ethical norms. But this overlooks that many, perhaps most of the natural properties that are expressed in the norm-condition of the general *prima facie* norms are *gradual*[13] and that these considerations in addition may come about in different *probabilities*. An example of the former type is the concept 'harming' (as it figures in one of Beauchamp and Childress' four principles) which may vary from killing a human being to inflicting some pain due to a necessary medical treatment. An example of the latter type is when '[a] pregnant woman has a serious heart disease that will probably result in her death if she attempts to carry the pregnancy to term' (Beauchamp and Childress 2001:129). Clearly, the degrees of both these aspects may have a say concerning the weights ascribed to the considerations at issue.

An additional point seems to be that the more general a *prima facie* norms is – i.e. the lighter (or wider) the norm condition of this norm is (e.g. *harming* as compared to *killing*) – the lesser weight it generally ascribes to cases which subsume under it.

11.13   The Application of this to the Moral Status of a Human Embryo

How does this approach to ethical justification come to terms with the moral status of human embryos and hence the morality of their potential use in medicine? We have earlier seen that a clear epistemic split between a normative and a descriptive

---

[13] All examples of thick concepts previously quoted from Williams clearly are gradual.

premise raises the problem how to justify the normative premise. Two alternatives of this type were considered – i.e. the metaphysical argument and the potentiality argument – neither of which seems to yield convincing answers. A better approach seems to be to accept a mutual epistemic dependence of the descriptive and the normative. But this dependence can not only be between the premises - as White apparently thought - but has to go a step further and include concepts. This points towards thick concepts – possibly in a somewhat wider sense than in Williams' understanding.

A crucial concept of this type is the natural property 'human being', as it figures, say, in the general norm 'Any killing of a human being is *prima facie* wrong' (and in the previously discussed arguments). This is a reasonable norm with a strong intuitive appeal, but raises several questions when applied: What is the *extension* of the concept 'human being' (which, qua a thick concept, is co-determined by ethical considerations)? Is this an *either-or* type (as the concept 'human embryo' clearly is from conception), or a *gradual* type (as 'grown up human being' equally apparently is)?

A further problem, as the previous example of *harming* versus *killing* suggests, is that the greater the extension ascribed to the concept 'human being', i.e. the further back in the pregnancy a human being is considered to have its beginning – the lesser intrinsic value does this apparently ascribe to the entity. A consequence of this is that some *other*, conflicting ethical concerns (e.g. the medical advantage of using human embryos) may outweigh this. The explanation for this is that the strength of our intuitions concerning the wrongness of killing a human being does not go all the way back to conception but increases throughout pregnancy. The killing of an embryo does hardly give rise to our feelings, while killing a baby, and even a fetus, does.

This, however, is not in itself a counter argument against the full moral status of the human embryo from conception, but rather what Rawls would call 'existing judgments' (Rawls 1971:48) or perhaps even 'provisional fixed points' (Ibid.:20). These 'existing judgments' are hence only the *outset* of a process towards a reflective equilibrium. And the outcome of this process may differ from its outset. The crucial question is whether this process will modify our 'existing judgments' about the extension of 'human being' and end with accepting human embryos as human beings. A possible reason for such a modification is if one 'can find an explanation for the deviation which undermines [one's] confidence in [one's] original judgments.' (Ibid.:48). The previously discussed metaphysical and potentiality arguments can be understood as attempts to 'undermine' this 'original judgment'. But if the previous criticisms of these arguments do hold water, they fail to do this. And if so the 'existing judgments' concerning the moral standing of human embryos survives the process towards a reflective equilibrium.

The conclusion is that one has to go for a gradualist understanding of the moral status of the embryo, i.e. a gradual increase in moral status towards a full moral status some time later in the pregnancy. This is gradual in a double sense: both the biological development of the fetus and the corresponding increase in moral status. This, of course, does not in itself give a precise answer to the question of the moral standing of the embryo, and does, hence, not tell what is permissible to do and at what stages of its development. To solve such problems one is left with one's judgments based upon intuitions. This, I take it, will shelter the use of embryos as much as possible, especially if there are alternative routes with similar prospects of outcome. But it will not block medical use of embryos – especially of 'supernumerary' embryos which will be destroyed anyway – if this is the most promising alternative in obtaining treatments of serious conditions which by now are without therapeutic treatments. Parkinson's decease is just one example.

11.14 Graduality and/or Created in the Image of God?

The gradualist view may in the outset seem incompatible with the metaphysical view that we as human beings are created in the image of God. This creation, some might claim, can not be the outcome of a gradual process: an entity is either created in the image of God or not.

But this seems to overlook the biological fact that human beings is the result of a gradual, evolutionary process. To accept this and still insist upon a full moral standing from conception seems to represent one of the following alternatives. The first rejects any graduality concerning moral status of humans, whether during the evolutionary history or during pregnancy. It hence accepts that at some definite, but unknown, point during the evolutionary history the first human being,[14] qua a creation in the image of God, was created in the uterus of an animal (i.e. of one of our non-human ancestors). This embryo was hence a human being with a full moral status while its parents were not since being animals and hence not created in the image of God. A consequence of this is that this embryo had a definite higher moral status than its parents. – This seems counter-intuitive. I do not see what kind of arguments could support this view.

The second, more reasonable alternative distinguishes between the evolutionary development and the embryonic growth during pregnancy and accepts a gradual increase in the moral status in the former sense but not in the latter. But since this, hence, in principle accepts a gradualist view, i.e. that there have been a growing moral status of the entities during the evolutionary process – a view which clearly is incompatible with an general either-or sense concerning moral status – it is difficult to see why not also accepting a gradualist view during pregnancy, even though only the latter type of graduality raises *present* moral problems.

The alternative to these views is to accept that being created in the image of God is compatible with a gradualist view in both these senses, i.e. an increase in moral status both during the evolutionary history and during pregnancy.

---

[14] Or rather two – a male and a female.

References

Beauchamp, Tom L. and Childress, James F (2001). *Principles of Biomedical Ethics*, 5th ed., Oxford: Oxford University Press.

Byrne, Peter (1988). "The animation tradition in the light of contemporary philosophy". In: Dunstan and Seller (eds.), *The Status of the Human Embryo. Perspectives From Moral Tradition*, Oxford: Oxford University Press.

Farley, Margaret A. (2001). "Roman catholic views on research involving human embryonic stem cells". In: Holland, Lebacqz and Zoloth (eds.), *The Human Embryonic Stem Cell Debate. Science, Ethics, and Public Policy*, Cambridge, MA: MIT.

Gärdenfors, Peter (1990). "Induction, Conceptual Space and AI", *Philosophy of Science* 57, Chicago: University of Chicago Press.

Quine, Willard van Orman (1960). *Word and Object*, Cambridge, MA: MIT.

Rawls, John (1971). *A Theory of Justice*, Cambridge, MA: Harvard University Press.

Singer, Peter (1979). *Practical Ethics*, Cambridge: Cambridge University Press.

White, Morton (1981). *What is and What Ought to be done. An Essay on Ethics and Epistemology*, Oxford: Oxford University Press.

Williams, Bernard (1973). "A Critique of Utilitarianism". In: Smart, J.J.C. and Williams, B. (eds.), *Utilitarianism: For and Against,* Cambridge: Cambridge University Press.

Williams, Bernard (1985). *Ethics and the Limits of Philosophy*, Cambridge, MA: Harvard University Press.

Williams, Bernard (1995). "Which slopes are slippery?" In: Bernard Williams (ed.): *Making Sense of Humanity and Other Philosophical Papers*, Cambridge: Cambridge University Press.

Chapter 12
The Beginning of Individual Human Life

Anthony Kenny

Abstract In this chapter the author asks the question when any individual life begins – at conception, at birth or at some point in between. He gives a survey of some of the ancient answers to this problem within Greek philosophy, in rabbinic texts and among Christian thinkers of antiquity and the Middle Ages. He also discusses alternative ways of phrasing that kind of question, including formulations in terms of animation or personhood. With regard to the moral status of embryos he argues that an embryo in the early days after fertilization cannot be seen as an individual human being because it may split into identical twins. He then refers to the Warnock Committee that offered a *terminus ante quem* for the origin of individual human life, namely the 14th day. He also himself supports placing the individuation of a human being somewhere around this point of time. And he concludes: 'But if the embryo, in its earliest days, is not yet an individual human being, then it need not necessarily be immoral to sacrifice it to the greater good of actual human beings who wish to conceive a child or reap the benefits of medical research'.

Keywords Human being, human life, individual/individuation, person(ality)/personhood, soul

When did I begin? When does any individual human being begin? At what stage of its development does a human organism become entitled to the moral and legal protection which we give to the life of human adults? Is it at conception, or at birth, or somewhere between the two?

The three alternatives – at conception, at birth, or between – do not in fact exhaust the possibilities. Plato, and some Jewish and Christian admirers of Plato, thought that individual human persons existed as souls before the conception of the

Faculty of Philosophy, University of Oxford, 10 Merton St, Oxford, OX1 4JJ, Great Britain, UK.
e-mail: ajpk@f2s.com

L. Østnor (ed.), *Stem Cells, Human Embryos and Ethics: Interdisciplinary Perspectives.*   167

bodies they would eventually inhabit. This idea found expression in the Book of Wisdom, where Solomon says 'I was a boy of happy disposition: I had received a good soul as my lot, or rather, being good, I had entered an undefiled body.' Clement of Alexandria records an early Christian notion that the soul is introduced by an angel into a suitably purified womb. Surely such fantasies have little relevance to any contemporary moral debate.

But in addition to those who thought that the individual soul existed before conception, there have been those who thought that the individual body existed before conception, in the shape of the father's semen. Onan, in Genesis, spilt his seed on the ground; in Jewish tradition this was seen not only as a form of sexual pollution but an offence against life. Aquinas, in the *Summa contra gentiles*, in a chapter on 'the disordered emission of semen' treats both masturbation and contraception as a crime against humanity, second only to homicide. Such a view is natural in the context of a biological belief that only the male gamete provides the active element in conception, so that the sperm is an early stage of the very same individual as eventually comes to birth. Masturbation is then the same kind of thing, on a minor scale, as the exposure of an infant. The high point of this line of thinking was the bull *Effraenatam* of Pope Sixtus V (1588) which imposed an excommunication, revocable only by the Pope himself, on all forms of contraception as well as on abortion. But the view that masturbation is a poor man's homicide cannot survive the knowledge that both male and female gametes contribute equally to the genetic constitution of the offspring.

At the other extreme are those who maintain that it is not until some time after birth that human rights arise. In pagan antiquity infanticide was very broadly accepted. No sharp line was drawn between infanticide and abortion, and as a method of population control abortion was sometimes regarded as inferior to infanticide, since it did not distinguish between healthy and unhealthy offspring.

In our own time a number of secular philosophers have been prepared to defend infanticide of severely deformed and disabled children. They have based their position on a theory of personality that goes back to John Locke. Only persons have rights, and not every human being is a person: only one who, as Locke puts it, 'has reason and reflection, and considers itself as itself, the same thinking thing, in different times and different places'. Very young infants clearly do not possess this degree of self-awareness, and hence, it is argued, they are not persons and do not have an inviolable right to life.

Defenders of infanticide are still, mercifully, very few in number. It is more common for moralists to take the rejection of infanticide as a starting point for the evaluation of other positions. Any argument that is used to justify abortion, or IVF, or stem-cell research must undergo the following test: would the same argument justify infanticide? If so, then it must be rejected.

The central issue, then, is to record, and decide between, the three alternatives from which we began: should we take individual human life as beginning at conception, at birth, or at some point in between? If the correct alternative is the third one, then we must ask further questions. What, in the course of pregnancy, is the crucial moment? Is it the point of formation (when the foetus has acquired distinct organs), or is it the point of quickening (when the movements of the foetus are perceptible to the mother)? Can we identify the moment by specifying a number of days from the beginning of pregnancy?

Some familiar texts from the Bible suggest that we should opt for conception as the beginning of the individual life of the person. 'In sin did my mother conceived me' sang the Psalmist (51,5). Job cursed not only the day on which he was born but also 'the night that said "there is a man-child conceived"' (3,3). Since 1869 it has been the dominant position among Roman Catholics, but for most of the history of the Catholic Church it was a minority view.

It has been much less common to regard personality and human rights as beginning only at the moment of birth. But one important rabbinic text allows abortion up to, but not including, the time when a child's head has emerged from the womb. Some Stoics seem to have taught that the human soul was received when a baby drew its first breath, just as it departs when a man draws his last breath.

Through most of the history of Western Europe, however, the majority opinion has been that individual human life begins at some time after conception and before birth. In the terminology that for centuries seemed most natural, the 'ensoulment' of the individual could be dated at a certain period after the intercourse that produced the offspring. Among Christian thinkers the general consensus was that the human soul was directly created by God and that it was infused into the embryo when the form of the body was completed which was generally held to occur around 40 days after conception.

Thomas Aquinas held a particularly complicated version of this consensus position. He did not believe that individual human life began at conception; the developing human fetus, for him, does not count as a human being until it possessed a human soul, and that does not happen until some way into pregnancy. For him the first substance independent of the mother is the embryo living a plant life with a vegetative soul. This vegetable substance disappears and is succeeded by a substance with an animal soul, capable of nutrition and sensation. Only at an advanced stage is the rational soul infused by God, turning this animal into a human being. Early term abortion, therefore, though immoral on other grounds, was not murder.

The whole process of development, according to Aquinas, is supervised by the father's semen, which he believed to remain present and active throughout the first 40 days of pregnancy. For this biological narrative Aquinas claimed, on slender grounds, the authority of Aristotle. At this distance of time, it is difficult to see why Aquinas' teaching on this topic should be accorded great respect.

A survey of the history of the topic makes it abundantly clear that there is no such thing as *the* Christian consensus on the timing of the origin of the human individual.[1] There was, indeed, a consensus among all denominations until well into the twentieth century that abortion was sinful, and that late abortion was homicide.

---

[1] Such a survey has been carried out with great care by David Albert Jones in his book *The Soul of the Embryo* (2004), to which I am greatly indebted for much historical information.

There was no agreement whether early abortion was homicide. However those who deny that it was still regarded it as wrong because it was the destruction of a potential, if not an actual, human individual. There was again no agreement whether the wrongfulness of early abortion carried over into the destruction of semen prior to any conception. Even within the Roman Church, different Popes can be cited in support of each option.

The question at issue is often posed in the confused form 'When does life begin?' If this means 'At what stage of the process between conception and birth are we dealing with living matter?' the answer is obvious: at every stage. At fertilization egg and sperm unite to form a single cell: that is a living cell just as the egg and sperm were themselves alive before their fusion. But this is clearly not the question which is relevant to the moral status of the embryo: worms and rosebuds are equally indubitably alive, but no one seeks to give their lives the protection of the law.

So perhaps we should reformulate the question: 'When does human life begin?' Here too the answer is obvious but inadequate: the newly formed conceptus is a *human* conceptus, not a canine or leonine one, so in that sense its life is a human life. But equally the sperm and the ovum from which the conceptus originated were human sperm and human ovum; but no one nowadays wishes to describe them as human beings or unborn children. If asked 'When does life begin?' We must respond with another question 'When does the life *of what* begin?'

Sometimes the question is formulated not in terms of life, but in terms of animation or personhood. Thus it is asked 'When does the soul enter the body?' or 'When does an embryo become a human person?' Modern discussions of the morality of abortion or of the status of the embryo often shy away from these questions, regarding them as matters of theology or metaphysics. Thus the Warnock Committee, whose report on human fertilization and embryology paved the way in England for the legalisation of experimentation on embryos, observed that some people though that if it could be decided when an embryo became a person it would also be decided when it might, or might not, be permissible for scientific research to be undertaken upon embryos. The committee did not agree.

> Although the questions of when life or personhood begin appear to be questions of fact susceptible of straightforward answers, we hold that the answers to such questions in fact are complex amalgams of factual and moral judgements. Instead of trying to answer these questions directly we have therefore gone straight to the question of *how it is right to treat the human embryo.*

A philosopher writing on these matters cannot evade, as a politician or a committee may do, the question of personhood. It is indeed a metaphysical question when personhood begins: that does not mean that it is an unanswerable question, but that it is a question for the metaphysician to answer. To answer it we must deploy concepts that are fundamental to our thinking over a wide range of disciplines, such as those of actuality and potentiality, identity and individuation: and these are the subject matter of metaphysics.

The question about personhood is also the same as the question about life, rightly understood. For 'When does life begin?' must mean 'When does the life of the individual person begin?'

The question is a philosophical one, but in order to answer it one does not need to appeal to any elaborate philosophical system, or appeal to quasi-theological concepts such as the soul. As so often in philosophical perplexity what is needed is not recondite information, or elaborate technicalities, but refection on truths which are obvious and for that reason easily overlooked.

If a mother looks at her daughter, six months off her 21st birthday, she can say with truth 'If I had had an abortion 21 years ago today, I would have killed you'. Each of us, looking back to the date of our birthday, can say with truth 'If my mother had had an abortion six months before that date, I would have been killed'. Truths of this kind are obvious, and can be formulated without any philosophical technicality, and involved no smuggled moral judgements.

Taking this as our starting point, however, it is easier to find our way through the moral maze. Let us consider first foetuses, and then embryos. Those who defend abortion on the grounds that foetuses are not human beings or human persons are arguing, in effect, that they are not members of the same moral community as ourselves. But truths of the kind that we have just illustrated show that foetuses are identical with the adult humans that are the prime examples of members of the moral community.

It is true that a foetus cannot yet engage in moral thinking or the rational judgement of action that enables adults to interrelate morally with each other. But neither can a young child or baby, but this temporary inability does not give us the right to take the life of a child or a baby. It is the long-term capacity for rationality that makes us accord to the child the same moral protection as the adult, and which should make us accord the like respect to whatever has the same long-term capacity, even before birth.

To be sure, there can be goodness of badness in human actions with regard to beings that are not members of the human moral community. Those who believe in God do not think of him as on terms of moral equality with us, and yet regard humans as having a duty towards him of worship. None-human animals are not part of our moral community, and yet it is wrong to be cruel to them. But the moral respect we accord to children, and, if I am right, should accord to foetuses, is something quite different to the circumspection proper in our relation with animals. For the individual that is now a foetus or a child will, if all goes well, take its place with us, as the animal never will, as an equal member of the moral community. As Kant might say, it will become a fellow-legislator in the kingdom of ends.

I have claimed it as an obvious truth that a foetus six months from term is the same individual as the human child and adult into which, in the natural course of events, it will grow after birth. This seems true in exactly the same sense as it is true that the child is the same individual as the adult into which it will grow, all being well, after adolescence. But if we trace the history of the individual from the foetus back towards conception, then matters cease to be similarly obvious.

Many people do not seem aware of the difficulty here. In 1985 in the UK the report of a committee chaired by Mary Warnock recommended the legalisation of experimentation on pre-implantation embryos; the committee's recommendation was put into effect by the Human Embryology and Fertilization Act of 1990. In a parliamentary debate triggered by the report of the Warnock Committee one Member of Parliament, Sir Gerald Vaughan, had this to say in opposition to experimentation on embryos.

> It is unarguable that at the point of fertilization something occurs which is not present in the sperm or the unfertilized ovum. What occurs is the potential for human life – not for life in general, but life for a specific person. That fertilized ovum carries the structure of a specific human being – the height, the colour, the colour of his or her eyes, and all the other details of a specific person. I do not think that there can be any argument against that. The fact that the embryo at that stage does not bear a human form seems to me to beg the issue and to be quite irrelevant. It carries the potential, and, just as the child is to the adult human, so the embryo must be to the child.

Sir Gerald concluded that human rights were applicable to an embryo from the first moment of conception.

It is undoubtedly true that in the conceptus there is the blueprint for 'the structure of a specific human being'. But to establish the conclusion that an embryo has full human rights, a different premise is needed, namely, that the conceptus contains the structure of an individual human being. A specific human being is not an individual human being. This is an instance of a very general point about the difference between specification and individuation. The general point is that nothing is ever individuated merely by a specification of its properties, however detailed or complete this may be – as it is in the case of the DNA of an embryo. It is always at least logically possible that there should be two or more individuals answering to the same specification; any blueprint may be used more than once. Two peas in a pod may be as alike as you please: what makes them two individuals rather than one is that they are two parcels of matter, not necessarily that they differ in description.

In the case of human beings the possibility of two individuals answering to the same specification is not just a logical possibility: it is a possibility that is realised in the case of identical twins. For this reason an embryo in the early days after fertilization cannot be regarded as an individual human being. The single cell after fusion is totipotential, in the sense that from it develop all the different tissues and organs of the human body, as well as the tissues that become the placenta. In its early days a single embryo may turn into something that is not a human being at all, or something that is one human being, or something that is two people or more.

It is important to be on guard here against an ambiguity in the word 'identical': there is a difference between specific identity and individual identity. Two things may be identical in the sense that they answer to the same specification, and yet not be identical in the sense that they are two separate things, not a single thing. When we say that Peter and Paul are identical twins we mean that they are alike in every specific respect, not that they are a single individual.

Between an embryo and an adult there is not an interrupted history of a single individual life, as there is linking foetal life with the present life of an adult. There is indeed an uninterrupted history of development from conception to adult; but there is equally an uninterrupted history of development back from the adult to the origination of each of the gametes that fussed at conception. But this is not the uninterrupted history of an individual. For each of the gametes might, in different circumstances, have fused to form a single conceptus, and the conceptus might, in different circumstances, have turned into more or less than the single individual that it did in fact turn into.

Of course, all development, if it is to proceed, depends on factors in the environment: an adult may die if diseased and a child may die if not nourished, just as an ovum will die if not fertilized and an embryo will die if not implanted. But though children and adults may die, they cannot become part of something else or turn into someone else. Foetus, child, and adult have a continuous *individual* development that gamete and embryo do not have.

The moral status of the embryo and the question whether its destruction is homicide was and is important, because if it is not, then the rights and interests of human beings may legitimately be allowed to override the protection that by common consent should in normal circumstances be extended to the early embryo. The preservation of the life of the mother, the fertilization of otherwise barren couples, and the furthering of medical research may all, it may be argued, provide reasons to override the embryo's protected status.

The line of argument I have outlined was found convincing in the United Kingdom not only by the Warnock committee but also by the later Harries committee.[2] These committees made a significant contribution to the debate by offering a *terminus ante quem* for the origin of individual human life – one which was much earlier in pregnancy than the 40 days set by the pre-reformation Christian consensus. Experimentation on embryos, they thought, should be impermissible after the 14th day. The Warnock committee's reasons were well summarized in the House of Commons by the then Secretary of State for Health, the Rt Hon Kenneth Clarke.

> A cell that will become a human being – an embryo or conceptus – will do so within fourteen days. If it is not implanted within fourteen days it will never have a birth [...] The basis for the fourteen day limit was that it related to the stage of implantation which I have just described, and to the stage at which it is still uncertain whether an embryo will divide into one or more individuals, and thus up to the stage before true individual development has begun. Up to fourteen days that embryo could become one person, two people, or even more.
>
> (Hansard 1985, vol. 73, col. 686)

This ethical reasoning is rejected by those Catholics who insist that individual human life begins at conception. An embryo, from the first moment of its existence, has the potential to become a rational human being, and therefore should be allotted full human rights. To be sure, an embryo cannot think or reason or exhibit any of the other activities that define rationality: but neither can a new born baby. The protection that we afford to infants shows that we accept that it is potentiality, rather than actuality, that determines the conferment of human rights.

---

[2] *Report of the Committee of Inquiry into Human Fertilization and Embryology* (chaired M. Warnock), 1984; *Report of the House of Lords Select Committee on Stem Cell Research* (chaired R. Harries), 2002.

Undoubtedly, whatever Aquinas may have thought, there is an uninterrupted history of development linking conception with the eventual life of the adult. However, the line of development from conception to fetal life is not the uninterrupted history *of an individual*. In its early days, as Kenneth Clarke indicated, a single zygote may turn into something that is not a human being at all, or something that is one human being, or something that is two people or more. Fetus, child and adult have a continuous individual development which gamete and zygote do not have. To count embryos is not the same as to count human beings, and in the case of twinning there will be two different human individuals each of whom will be able to trace their life story back to the same embryo, but neither of whom will be the same individual as that embryo.

Those who argue for conception as the moment of origin, stress that before fertilization we have two entities (two different gametes) and after it we have a single one (one zygote). A moment at which one entity (a single embryo) splits into two entities (two identical twins) is surely equally entitled to be regarded as a defining moment. It is true that in the vast majority of cases twinning does not actually take place; but surely the strongest element in the Catholic position is the emphasis it places on the ethical importance of potentiality. It is the potentiality of twinning, not its actuality, that gives reason for doubting that an early embryo is an individual human being.

In my view, the balance of the arguments lead us to place the individuation of the human being somewhere around the 14th day of pregnancy. But there are two sides to the reasoning that leads to that conclusion. If the course of development of the embryo gives good reason to believe that before the 14th day it is not an individual human being, it gives equally good reason to believe that after that time it *is* an individual human being. If so, then late abortion is indeed homicide – and abortion becomes 'late' at an earlier date than was ever dreamt of by Aquinas.

Since most abortion in practice takes place well after the stage at which the embryo has become an individual human being, it may seem that the philosophical and theological argument about the moment of ensoulment has little practical moral relevance. That is not so. If the life of an individual human being begins at conception, then all practices which involve the deliberate destruction of embryos, as whatever stage, deserve condemnation. That is why there has been official Catholic opposition to various forms of IVF and to scientific research involving stem cells. But if the embryo, in its earliest days, is not yet an individual human being, then it need not necessarily be immoral to sacrifice it to the greater good of actual human beings who wish to conceive a child or reap the benefits of medical research.

References

Jones D.A. (2004). *The soul of the embryo: An enquiry into the status of the human embryo in the Christian tradition.* London/New York, Continuum.
Report of the Committee of Inquiry into Human Fertilization and Embryology (chaired M. Warnock). London, HMSO, 1984.

Report of the House of Lords Select Committee on Stem Cell Research (chaired R. Harries). London, HMSO, 2002.
Thomas Aquinas (1975). *Summa contra gentiles*. Notre Dame, IN, University of Notre Dame Press.

Chapter 13
Embryonic Stem Cell Research – Arguments of the Ethical Debate in Germany

Ludger Honnefelder

Abstract Central to ethical debate on embryonic stem cell research in Germany are on the one hand visions in the life sciences and medicine of understanding the early differentiation processes of the human embryo as well as the development of novel therapeutic strategies for otherwise incurable diseases by using ES cells and, on the other hand, fundamental ethical concerns regarding the destruction of human embryos required for the generation of ES cells. The answer to the question whether it is ethically acceptable to use ES cells in research depends basically on the question of the *moral status of the human embryo*, i.e. what good the embryo represents in terms of a moral assessment and what level of protection ought to be granted. Although German legislation, particularly the Embryo Protection Act, provides clear-cut principles on this question the ongoing debate in Germany reveals substantial controversies in the assessment of the moral status especially of those embryos which were generated in vitro and are in the early stages of development, i.e. prior to the formation of the primitive streak and nidation in the mucosa of the uterus. As in other western industrial countries basically *two positions* on the moral status of the embryo – a restrictive and a gradualist position – can be distinguished. It is the aim of the present paper to analyze these two positions in the context of the ethical debate in Germany.

Keywords Stem cell research, embryo, moral status, restrictive position, gradualist position

13.1   Introduction

Central to ethical debate on embryonic stem cell research in Germany are on the one hand visions in the life sciences and medicine of understanding the early differentiation processes of the human embryo as well as the development of novel

Institute of Science and Ethics, Bonner Talweg 57, 53113 Bonn, Germany.
e-mail: honnefelder@iwe.uni-bonn.de

therapeutic strategies for otherwise incurable diseases by using ES cells and, on the other hand, fundamental ethical concerns regarding the destruction of human embryos required for the generation of ES cells. To do justice to these two – partially conflicting – aspects represents an ethical challenge. For if the insights into the developmental mechanisms of early human life and the cure of human life threatened with disease cannot be achieved in any other way but the generation and subsequent destruction of new human life, stem cell research confronts us basically with the moral decision of whether and – if so – to which extend it is justified to sacrifice human life for other human life.

The ethical problems associated with the sourcing and the use of ES cells from human embryos can be summarized by two key questions: (1) To what extent is medical research necessarily relying on embryonic stem cells derived from human embryos? And (2) what is the level of protection we ought to grant to the human embryo?

13.2   Aims and Means of Research with ES Cell Lines

Underlying the first question is an acknowledgement that one cannot avoid a careful consideration of the ethical values inherent in the aims and means of research with embryonic stem cells. With regard to its aims two main fields of research can be identified at present: *basic research* and *applied medical research.*

Basic research includes the scientific examination of cell differentiation, tissue organization as well as the search for biological molecules essential for the reprogramming of differentiated cells into earlier stages of development, all of which is difficult to study in the intact human embryo. Other topics in basic research include the development of diagnostic tools for the in vitro screening of desired and undesired effects of drug candidates and for toxicological studies in cells differentiated from ES cell lines into tissue type specific cells. This type of research may result in an increase in drug safety as well as support food and environmental chemistry by substantially accelerating and widening the possibilities of pre-clinical drug testing.

The second field of research using ES cells, i.e. applied medical research, predominantly aims at developing transplantation therapies for diseases which, at present, can barely be treated at all due to a lack of appropriate drugs. Furthermore, using ES cells for transplantation the improvement of therapeutic strategies is envisioned in diseases where available treatment is normally sufficient but associated with certain inconveniences for the patients, such as in insulin-dependent Diabetes mellitus.

It is obvious that in an ethical assessment both, basic research and applied medical research cannot be entirely separated from each other. Yet the question arises as to the criteria that should apply to the moral assessment of the aims of ES cell research. Does it make an ethically relevant difference whether research using ES cells focuses only on the development of therapies or whether aims which are not therapy-related are also pursued? Is it morally justified to focus research with ES cells only on diseases which are not treatable at present or should this type of research cover all areas with potential therapeutic implications? Is it relevant for the moral assessment that – by now – it is not at all certain whether research using ES cells will result in any useful outcome at all?

With regard to the aims of research the fundamental ethical regulations behind the German constitution, the Basic Law, guarantee the independence of research from any kind of political and social restrictions. The *freedom of research and the sciences*, as established in Article 5, Paragraph 3 of the Basic Law, follows from the protection of human dignity guaranteed in Article 1. It can only be restricted if and where it conflicts with other constitutional guarantees, such as the right to life and to inviolability of the person (Article 2, Paragraph 2). It follows that within the context provided by the German Basic Law the aim to find treatments for diseases and ways to prevent them, but also the general goal of gaining knowledge must be regarded as morally high ranking. With regard to the goals of research involving human ES cells both principles, the right to life and the freedom of research and sciences, are not in conflict. Yet, as the sourcing of ES cells requires the destruction of early human life, the question arises whether a situation of conflict occurs with respect to the means of ES cell research and, if so, what criteria may allow an ethically justified decision. This question requires a reflection on the ES cells used as means in research.

13.3 The Restrictive and the Gradualist Position

The answer to the question whether it is ethically acceptable to use ES cells in research depends basically on the question of the *moral status of the human embryo*, i.e. what good the embryo represents in terms of a moral assessment and what level of protection ought to be granted. Although German legislation, particularly the Embryo Protection Act, provides clear-cut principles on this question the ongoing debate in Germany reveals substantial controversies in the assessment of the moral status especially of those embryos which were generated in vitro and are in the early stages of development, i.e. prior to the formation of the primitive streak and nidation in the mucosa of the uterus. As in other western industrial countries basically *two positions* on the moral status of the embryo can be distinguished. One position holds that, beginning with the existence as a single totipotent cell, the embryo ought to be acknowledged as a good which deserves unrestricted protection as it is granted to any born individual. The other position claims that the moral status of an early embryo differs from that of later stages and, accordingly, the level of protection to be granted increases as certain stages of development are reached.

For proponents of the first – *restrictive* – position, the notion of human dignity – which is founded in the capability of the human being to be subject of and responsible for his acts and thus being conceivable as an end in itself – applies to the early embryo since right from the beginning: from this time on the embryo has the potential to develop into a moral subject and the embryo and the moral subject are identical, i.e. the same human being. The *identity* of the moral subject with the embryo corresponds to the *continuity* in the development of the embryo, which does not allow for the identification of certain developmental stages as a basis for the moral assessment of the embryo's status. Since the dignity of the moral subject is entitled to protection, the two notions of identity and continuity need to entail the same protection for any early stage of human development which in itself bears the potentiality to develop into a moral subject. This potentiality is already present in a single cell stage embryo when the individual genome directing the development of the human embryo is constituted. As a consequence, proponents of this position call for the full protection of the embryo starting from the earliest beginnings of life.

Others however, while still acknowledging human dignity attributed to the embryo in the totipotent single cell stage, do not hold such a restrictive position with regard to the protection of life. They differentiate between dignity and the protection of life claiming that, under certain conditions – as, for example, in the case of super-numerary embryos in vitro, who have no perspective of development – it is no viola-tion of human dignity if those embryos are denied the full protection of life.

Proponents of the second – *gradualist* – position claim that an embryo or fetus should only be granted the same level of protection as that applied to a moral sub-ject if certain stages of development or certain qualities characteristic of a moral subject have been reached. An extreme version of that gradualist approach, which is, however barely present in the German debate, considers the capability of having interests as one of those requirements and preconditions for a full protection of life. Accordingly, not even the moment of birth is considered as a precondition for full protection. More moderate versions of the gradualist view, however, regard certain conditions in the normal development of the human embryo to be relevant for the extent to which an embryo must be protected. Those conditions include the indi-viduation of an embryo, i.e. the loss of the early embryo's potential to divide into two or more embryos, the occurrence of heart action, the appearance of neurons, first movements of the embryo in the mother's womb as well as their perception by the mother, the nidation of the embryo into the uterus, and others.

Both, restrictive and gradualist approaches, the latter with the exception of the more extreme version of this position, have in common that they acknowledge the completed fertilization of a human oocyte as the beginning of life of a human being and that, starting from this point, the moral status of the embryo is to be judged in relation to the status applying to a moral subject. Furthermore, both positions agree on a view of early human life as a good that must not be utilized at will but ought to be granted at least a certain level of protection. Thus, both positions have some common ground in understanding early human life as a good which has a value independent of the approval of other individuals and, therefore, ought to be protected.

Yet, the differences between the two positions may result in substantially different consequences with regard to the actual protection of the embryo's life when, faced with competing goods, a decision has to be made on the preference as it may be the case regarding research with ES cells. From a restrictive point of view, *weighing the protection of the embryo against other goods* is considered to be either excluded per se or legitimate only in the context of a conflict relating to the most highly valued goods or inevitable evil. For proponents of the gradualist position, however, weighing the protection of an early embryo against other goods is considered to be justifiable since – in that view – the embryo does not deserve full protection at an early stage of development. Nonetheless, as the embryo is viewed as a high-ranking good in itself, weighing is justified only if the competing good represents a high ranking value as well.

13.4   Ethical Assessment of the Different Ways of Generating Embryos Used for the Isolation of ES Cells

In addition to the assessment of the moral status of the human embryo, the decision on whether or not to use ES cells as a means for research purposes has to consider the *intentions and the mode of generating embryos* used for the isolation of ES cells. Even some proponents of the gradualist position see human dignity violated by the *generation of embryos exclusively for research*, as the notion of human dignity implies that human life has intrinsic value independent of the approval and aims of others.

Moreover, some views represented in the German debate consider the generation of embryos by nuclear replacement technology as even more problematic in ethical terms. Deriving ES cell lines from those embryos for the purpose of transplantation therapy may result in the potential clinical benefit of reducing or entirely overcoming the immunological rejection of transplanted cells. However, not only may the generation of those embryos for the exclusive aim of research purposes be viewed as violating the notion of human dignity. A further question may also arise on whether the generation of an embryo by nuclear replacement technology constitute a new level of treating human life only as a means and not also as an end in itself, because the entire genetic makeup of that embryo is chosen and assigned to it by others and is identical with the genetic makeup of an already existing human individual.

Those ethical concerns with regard to the intentions and the mode of generating human embryos for research purposes do not apply to *supernumerary embryos*. These embryos were generated for the purpose of being born but this purpose had to be abandoned due to reasons on the part of the mother. Thus, as the adoption of in vitro embryos can apply only to individual cases, these embryos do not have any perspective of getting the chance to develop and be born. The question, then, arises whether it is justified to utilize those embryos for high ranking research purposes such as the development of therapies. For proponents of the restrictive position the decision concerns the question whether the destruction of an embryo for the sake of high ranking research goals violates the notion of human dignity, even given the circumstances that the embryo has no chance to develop anyway. For many of them this condition does not justify the destruction of early human life since they acknowledge an ethically relevant fundamental difference between the lack of a chance to develop and a deliberate destruction of the embryo's life for research purposes, the latter being considered an instrumentalization of the embryo which violates the notion of the human dignity. In contrast to this, proponents of a gradualist position consider it to be justified to prefer the value of the freedom of research over the protection of early human life since the destruction of such life is not viewed as an instrumentalization which violates the notion of human dignity.

13.5   The 'Protection-Worthiness' of the Embryo – Further Considerations

What follows from these reflections? Conferring a moral status upon a person only once he or she is born and otherwise merely speaking of pre-emptive protection for the unborn person cannot be a persuasive approach. For insofar as the arguments that have been set out do not fail on their own premises, they substantiate the granting of a moral status only long after birth and hence run contrary to the essential intuitions expressed by the idea of human rights. If the protection of human dignity is the entitlement of humans as humans, the commencement of such protection cannot be made dependent on any conditions other than the very commencement of 'being human'.

This *protection-worthiness* raises questions in relation to the early phase of the human embryo, especially in relation to the embryo in vitro. How is recognition of the protection-worthiness due to human beings from their inception to be reconciled with our understanding of the progressive nature of their creation and development? It is striking to note that the majority of those who argue for graduated protection in step with development and geared in particular to the time of nidation also do not wish to leave the prior period after completed fertilization without protection. Evidently, there is a *commonality between the discussed positions* of graduated and non-graduated protection that lies in a fundamental conviction that at no point in their existence can human beings be disposed of in any way desired and therefore they must in principle be regarded as worthy of protection.

The *differences* begin with the question of the extent to which the fundamental practical judgment that is associated with the use of the sortal expression human and that assigns humans as humans a moral status should be extended to the artificially created embryo – or, to put it another way: whether and if so how this judgment is to take account of the *progressive nature of embryonic development*. The difference in the answer provided to this question does not arise out of any disparity in the underlying facts. For in neither case is the practical judgment at issue here derived from biological or metaphysical facts. Rather, what is at stake is how moral relevance must be attributed to the empirical assumptions that are to be taken into account: in the one instance, the practical judgment is that even where uncertainty remains as to the inception of a human being the protection of human dignity demands inclusion of the embryo from the moment of completed fertilization onwards. In the other instance, the practical judgment tends towards only allowing this protection of dignity to attach from the point in time when the real potentiality has acquired its definitive weight through implantation in the mother's uterus, while only attributing derived protection to the embryo in vitro existing prior to this point in time.

A gradualist understanding faces the difficult question of how the moral relevance of the caesura invoked for gradation of the status can be demonstrated in light of the fact that the embryo undoubtedly already constitutes a human being prior to the date of implantation. What is more, it must account for how far the limited protection-worthiness of the embryo which it grants prior to this date can be substantiated in this case and whether such a limited protection-worthiness can satisfy the demands embodied in the idea of human rights. Indeed, can a protection-worthiness attributed to the embryo in vitro be understood in any other way than under the sanctity of the dignity to which humans in general are entitled? If we accept that the protection of dignity can have a pre-emptive effect akin to the after-effect in relation to the human corpse, either we are disputing that the early embryo is already a human being or the assumption of such protection-worthiness presupposes – like any anticipatory form of protection-worthiness – that what is to be protected falls under the sortal predicate human being. In this case, however, it is scarcely possible to justify a gradation within status or dignity. If we seek a way out by differentiating between a human dignity that cannot be weighed in the scales and one that can, we miss the point of the term 'human dignity' – which, after all, is precisely intended to finally and fundamentally prevent man from being weighed up against other things.

Conversely, the understanding that assigns an unlimited moral status to the human embryo from completed fertilization onwards and hence considers it subject to the protection of human dignity must face the objection that the development from fertilization to birth is a process and the determination of the moment at which human life begins therefore involves difficult questions. What is more, the question arises as to whether allowance can only be made for the aforementioned progressive nature of early embryonic development in the manner already discussed, i.e. by resorting to the assumption of a graduated dignity or a graduated protection-worthiness – with the associated difficulties – or whether one continues to assert the appropriateness of the protection of dignity for the embryo in vitro too and takes account of the progressive nature through allowance for *circumstances*.

This is particularly relevant to the correlation between the protection of human dignity and the protection of life. If we assume that protection-worthiness is graduated in such a way that the embryo in vitro enjoys only a derived protection, but not that of human dignity, the possibility arises of weighing this protection up in the face of high-order goals that cannot otherwise be attained. Indeed, the intuition that such a weighing up should be possible would appear not infrequently to be the reason behind the proposed solution that protection-worthiness should be correspondingly graduated.

The situation is different if one adopts the position of unlimited protection first referred to above. In this case the question arises as to whether and how the progressive nature of embryonic development can be a reason for considering a weighing up to be possible in view of the protection of life that follows from the protection of dignity. From the standpoint of the same unlimited protection of dignity a number of different answers to this question can be identified: starting out from the idea that the life of a human being constitutes the fundamental condition for the ability to be a subject, a first answer concludes that unlimited protection of life is to be considered imperative and that a limitation can only be permitted in the exceptional case that *life stands against life*. A second answer assumes that unlimited protection of dignity can only evolve into a corresponding protection of life depending on the circumstances. Consequently, a weighing up in the face of high-order goals that promote the protection of life is possible in the event that an embryo that has been created to bring about pregnancy cannot be used for this purpose for reasons that cannot be rectified and the protection of dignity can therefore only be realized by allowing it to die. In this understanding, it would not therefore constitute a violation of the required protection of dignity to remove stem cells from the embryo under these strictly defined circumstances, even if this leads to the destruction of the embryo that has been left to die. In this context, it is assumed that the requirement for the protection of life derived from the protection of dignity manifests itself differently on account of the circumstances described in this case compared, for example, to cases of infaust diseases or moribund states.

13.6   Outlook

The remaining questions – which are without doubt of weighty significance – can no longer be resolved purely through recourse to the moral status of the embryo. For such clarification necessitates a more intensive exploration of the question as to how the claims to protection deriving from the protection of dignity should evolve and whether – and if so to what extent – allowance can and must be made for conflicting circumstances that may exist in this context. Recourse to moral status is indispensable if we wish to be guided by the idea of human rights and wish to make man's intrinsic protection-worthiness contingent not on conferral by third parties, not on certain performances or qualities and not on scenarios that involve weighing up, but rather solely on the *recognition of humans as humans*, i.e. on the moral and legal status that is intrinsic to 'being human'. Insofar as we express this status in the value judgment associated with the use of the sortal predicate human and tie it to human beings, everything would support allowing it to begin with the life that is intrinsic to human beings.

References

Deutsches Referenzzentrum für Ethik in den Biowissenschaften (ed.) (2001). "Blickpunkt, Forschung mit humanen embryonalen Stammzellen", http:// www.drze.de/ themen/blickpunkt/ Stammzellen

Honnefelder, Ludger (2002). "Die Frage nach dem moralischen Status des menschlichen Embryos". In: O. Höffe, L. Honnefelder, J. Isensee, P. Kirchhof (eds.). *Gentechnik und Menschenwürde. An den Grenzen von Ethik und Recht*, DUMONT Literatur und Kunst Verlag, Köln, Germany, pp. 79–110.

Honnefelder, Ludger and Heinemann, Thomas (2002). "Principles of ethical decision making regarding embryonic stem cell research in Germany". In: R. Chedwick and U. Schüklenk (eds.). *Bioethics*, Special Issue: *Stem Cell Research*, Vol. 16/November 2002, Oxford, UK/ Boston, MA, pp. 530–543.

Heinemann, Thomas and Kersten, Jens (eds.) (2007). "Stammzellforschung. Naturwissenschaftliche, rechtliche und ethische Aspekte". *Sachstandsberichte Nr. 4 des Deutschen Referenzzentrums für Ethik in den Biowissenschaften,* Karl Alber Verlag, Freiburg, Germany.

Jahrbuch für Wissenschaft und Ethik 7 (2002). Contains a number of articles on the ethical question of stem cell research in Germany, Walter de Gruyter Verlag, Berlin, Germany, pp. 5–52, 165–195.

Chapter 14
The Question of Human Cloning in the Context of the Stem Cell Debate

Otfried Höffe

Abstract The debate on stem cell research is as heated as it is polarised. Sober analysis of the issues at hand is in order, an analysis which starts with defining three fallacies, each of which leads to a misrepresentation of certain aspects of the debate in consideration. Once these fallacies are identified, the analysis goes on to isolate the relevant features of stem cell research. On the technological side, research is currently still risky, as the biggest part of experiments fail. Even where they seem to succeed, as in the case of Dolly, negative implications abound: Dolly was not excessively vital and died prematurely. These normative objections on the technical side, however, are open to technical solutions. There are, on the other hand, normative aspects, that remain even if the technical problems are solved. Among these are the question of the status of the human embryo and its possible relativity in the realm of research cloning, a relativity that may collide with the moral imperative not to destroy human life even to save other human lives; on the side of reproductive cloning there always remains the question of self-esteem and esteem from others with respect to human beings that have been designed and produced by other human beings rather than biological luck. As a consequence, while it is hard to defend a categorical ban on human cloning, moral as well as technical objections form a strong case for a very cautious stance.

Keywords Argumentational fallacies, Kant, reproductive cloning, research cloning, stem cell debate

14.1 Introduction

If a long debate of experts has not lead to a reasonable consensus, it might be helpful to take a step backwards. As it is true for the debate on human embryos and stem cell research that it has not yet achieved sufficient agreement, I shall not enter

Philosophical Seminar, Eberhard Karls-University Tübingen, Bursagasse 1, 72070 Tübingen. Germany, e-mail: sekretariat.hoeffe@uni-tuebingen.de

L. Østnor (ed.), *Stem Cells, Human Embryos and Ethics: Interdisciplinary Perspectives.* 187
© Springer Science + Business Media B.V. 2008

immediately into the question of the moral status of human embryos but discuss questions which are both basic and preliminary. Only in the end shall I present and defend my cautious answer.[1]

I begin with the very basic remark that some biological and medical researchers fear that ethics, whether philosophical or theological, acts primarily as a border guard and veto authority. In reality, for the life sciences law elevates the freedom of research to the status of a fundamental right, which overrides ordinary rights. And philosophical ethics views the thirst for all knowledge, especially the one which is free of all utility, as the highest form of knowing. If research, by contrast, enters into human service, then the life sciences find themselves in the fortunate position of referring to the interculturally recognized duty to help which, in the case of the doctor, is part of his professional ethics, even of the Hippocratic oath. Human cloning does, of course, concern other fundamental, therefore overriding rights. Only because of them, law and ethics are called upon to act as border guards, with the result that there can be no legitimation without limitation. The Hippocratic oath states uncompromisingly: *primum nil nocere*, first and foremost do no harm.[2]

14.2 Three Argumentational Fallacies

Adequate reflection must avoid three mistakes. The fact that they are mistakes is obvious, but only in principle, whereas in concrete bioethical debates these fallacies are occasionally committed. The first two fallacies want to refuse the necessary cooperation of biology and ethics whereas the third one underestimates a condition which is not quite new but has reached a particular topicality.

The first, the is-ought fallacy, more specifically the 'biologistical fallacy', is committed by natural scientists if they wish to draw a demarcation line between the permitted and the non-permitted research only on the basis of biology. Whoever looks more closely will discover hidden normative parameters, e.g. the adaptability of our species. But anyone falls foul of the is-ought fallacy, too, who, the other way around, justifies research boundaries by saying that man is not the creator but a created entity. Though this remark might be right, a *de*scription of the human condition does not contain any potential for *pre*scriptions. This is also a problem when one turns new options into taboos by saying that man should 'not play God'. Because scientists are dependent on given materials, they are not on a par with divine *creatio ex nihilo* (creation out of nothing). By contrast, they are both capable and – within the framework of law and ethics – entitled to indulge in subdivine creation,

---

[1] This text is a substantially revised and enlarged version of my essay "Human cloning: The legal-ethical debate. An interim stocktaking", in: Honnefelder and Lanzerath (2003), pp. 453–463.

[2] For ethics in general cf. Höffe (2007b); for morals as a price of modernity cf. Höffe (2000); for biomedical ethics in general cf. Höffe (2003), references there; Geyer (2001); Burley (1999).

in post-creation respectively created co-creation as they are beings created in the image of God. And the fact that post-creations are imperfect is one of the reasons why biomedical research is a never-ending process. Even with respect to the species, the well known potentiality and the continuum arguments are in danger of falling into the is-ought-mistake.

The second fallacy, opposite to the is-ought one, the ought-is fallacy or 'moralistic fallacy', is committed by moralists who introduce their prohibitions without sufficient knowledge of the biological facts. For instance, they believe cloning is a purely artificial, man-made procedure and regard mere artificiality as a counterargument. In reality cloning already happens in nature: plants and simple animals clone, though because of the disadvantage of chromosome-identical copies and a general 'interest' in diversity, nature prefers sexual reproduction in the case of more complex, 'higher' animals.

The third bio-ethical fallacy consists in overestimating the justifying power of one's own moral convictions. This violates our political life-world, which exists within nations in a liberal, pluralistic democracy and, on a global scale, in an even more diverse co-existence of ideologies and religions. In this situation, it is not allowed to use as a final argument any non-universally valid principle, for instance one based on an obsolete metaphysics or religious premises. In a pluralistic, even multicultural moral and legal world, ethicists have to abstract from their personal moral or religions convictions and follow a secular line of argumentation. At least, they have to follow a modestly secular line which admits both, on the one hand, that there might be religious strategies of justification, but on the other hand that the religious part as such cannot be the decisive factor for universally valid solutions.

Fortunately, since its very beginning with Plato and Aristotle, philosophical ethics respectively moral philosophy merely appeals to the common reason of man and to common human experiences. Another advantage now for applied ethics in contrast to fundamental moral philosophy is that it does not seek an ultimate justification ('Letztbegründung'). It rather argues using a topical method which refers to widely recognized moral principles.[3]

In order to identify those principles we have to consider the decisive normative aspect. In the case of human cloning the debate is not directed at the personal question of whether a researcher may or must not clone, but whether the law shall permit or forbid it, sometimes with the additional question of whether a legally permitted research should be subsidized by tax payers. For this debate the widely, even universally recognized moral principles consist in the fundamental basic human rights. By combining biological facts with these legal moral principles, or more precisely by judging biological facts in the light of those principles, one avoids all three

---

[3] Cf. Aristotle, *Topics*, Chapter I and Rawls' concept of an overlapping consensus: Rawls (1993), lecture IV.

fallacies: The facts correspond to the is-part, the rights to the ought-part, and last not least they meet the requirement of intercultural recognition.[4]

14.3   Reproductive Cloning?

When new phenomena appear, they are soon joined by twins who, of course, have no genetic identity: hope and fear. For a long time they were nourished in biomedicine by opposing sides: the hopes of research scientists and the fears of the alarmed public. Fortunately, meanwhile this 'division of labour' has been overcome.

When research is undertaken without blinkers, the risks become obvious: Dolly the sheep was not excessively vital. Even in the model organism for mammals, the mouse, the well-known Rudolf Jaenisch, member of the Whitehead Institute for Biomedical Research at the Massachusetts Institute for Technology, reported that manipulations of egg cells only succeed once in every 104 attempts. Even in the case of successful manipulations, dozens of important genes in the transferred nucleus go off-course. And according to reproduction expert Gerald Schatten, the experiments with rhesus monkeys, which were consciously chosen as representatives of man, have all failed. The probable cause: already during the first cell divisions, the chromosomes did not align themselves correctly. Because of the lack of important proteins the so-called spindle apparatus was disorganized and as a consequence genetic disregulation was almost unavoidable.

Does this mean an overhasty hope has died and along with it the threat that ethics and law might become unemployed as border guards? There is now overwhelming approval for the ban on reproductive cloning, both amongst natural scientists and in ethics and law. Rudolf Jaenisch and the 'Dolly father' Ian Wilmut state in the journal *Science*: 'Don't Clone Humans!'[5] The US report on *Human Cloning and Human Dignity*[6] calls for a strict ban, for an unlimited period of time, on human cloning for the purposes of creating children. Five years earlier, the European Parliament, in November of 1997, and the General Assembly of the United Nations came out in favour of a ban on human cloning, in some cases a comprehensive ban, in others a ban on reproductive cloning.

Nevertheless, the philosopher prefers to be sure. If the arguments are not sound, the consensus will begin to crumble. After all, in the debate in the 1960s and 1970s there were also advocates[7] of human cloning who were joined ten years ago by almost two dozen high-ranking research scientists.

---

[4] A broad overview of the debate is offered by: German Reference Centre for Ethics in the Life Sciences (DRZE) (2003), Forum Diderot (1999), McLaren (2002), also The President's Council on Bioethics (2002), Chapters 5–6, Honnefelder and Lanzerath (2003).

[5] Jaenisch and Wilmut (2001).

[6] See note 4.

[7] For example, Fletcher (1972), critical Jonas (1972).

Though our main interest concerns the so-called therapeutic cloning, I will start by considering the reproductive cloning of humans. There are three obvious arguments for the comprehensive ban: firstly, already in the case of higher animals, cloning is not just enormously complicated but also very risky; by far, most of the experiments end in miscarriage or deformity. Secondly, in the case of humankind, even severely handicapped people have a full right to life. Thirdly, it would be dreadfully boring if everyone were the same; everybody wants to be different.

At first sight each of these arguments, and more particularly the combination of them, point to an absolute ban. At second glance, when it comes to the question of the scope of the arguments, only a relative ban remains. The middle, legal and moral argument is not disputed. It may even be joined by a civil law aspect which is, however, disputed: if a handicapped person is created through cloning, he has on the one hand a full right to life but can, on the other, claim legal damages himself or through a representative. It is this risk which makes people wary of reproductive cloning.

For the first argument, the fact that the procedure is 'enormously risky', there are indeed good biological reasons why one can, along with the Nobel prize winner Christiane Nüsslein-Volhard, talk of a 'resistance of nature' and also why the procedure can be described as 'highly unnatural'.[8] Nevertheless, it may be only a matter of time: the procedure might be at present, and perhaps for a long time to come, risky, but not necessarily 'forever' so. For today's legislator there is, of course, the clear requirement of the Hippocratic oath, the *nil nocere*, as normative background.

The alleged right to reproductive freedom, publicized as reproductive autonomy by the legal philosopher Ronald Dworkin,[9] ranks far lower than our responsibility towards the predictable (!) high risk to the well-being of the future child. It is true, an intimate area like reproduction demands a high degree of reticence from the legal order. A thought experiment, however, shows that absolute reticence, unrestricted reproductive freedom, would be absurd. Just imagine parents who endeavor – because of scarcely comprehensible cynicism – and who, in an even crasser form of cynicism, agree with others to give birth to what are probably going to be severely handicapped children: Should the legal order here really do nothing?

Highly risky cloning not only violates the principles of specific biomedical ethics but already those of the general ethics of human experiments. When, however, thanks to an enormous improvement in knowledge and ability, the risk of handicap becomes 'extremely low' – the cloning ban would lose its legal justification.

The third argument, again an assessment of facts, does not apply equally to all cultures and eras. But in our culture the desire to be special or different is probably the norm. Nevertheless, two small questions arise. Firstly, genetically identical human beings already occur naturally, as monozygotic twins, without foregoing the

---

[8] Nüsslein-Volhard (2003), 10.
[9] Dworkin (1996), 104 f.

prospect of being special. After all, human individuality cannot be reduced to a set of chromosomes. Even in the case when the genes are absolutely identical, as in monozygotic twins, one cannot reliably infer the traits of the one based on knowledge of those of the other. Besides lesser influences such as cytoplasm and pregnancy, a colourful array of cultural and social factors must be considered. Both sides play a main role: the individual and how he reacts to partly biological, partly social preconditions and how these preconditions are looked upon and what is done with them.

Obviously, this further undermines the justificatory potential for an absolute ban. It also, of course, undermines the other side, the parents' expectation. Parents are not so much interested in something abstract like a genetically identical offspring but far more, in a concrete way, in children who are particularly successful in certain manual and social, intellectual or artistic terms. There is absolutely no way that cloning could guarantee this. Nor can it in any way ensure that the desired abilities will be as highly esteemed in the future social environment both objectively by society and subjectively in the life plans of the concrete child. Above all, cloning does not guarantee the most important goal of responsible parents, that is, neither narcissism, nor over-ambitiousness, nor lust for power, but a certain degree of self-esteem of the children as well as the esteem for others, both forms of esteem that are required for a successful life. The dream of immortality is again utopian in the strictest sense of the 'nowhere'. After all, the donating individual 'dies' whereas the alternative, living on through one's children, does not require cloning. The same applies to deceased partners and children. They are far more than a genetic code whereby this "more", by definition, is to be denied to the clone.

We have to thank Immanuel Kant not only for the great term 'human dignity' but also for the pragmatic proposal of a thought experiment: in order to examine a judgement for more than its own validity, one should 'put oneself in the position of everyone else'.[10] Anyone who is considering cloning, must therefore ask himself if he would rather have been born as a clone, i.e. without the usual open-endedness with respect to (1) his image, (2) his life and future plans, (3) his self-esteem and (4) the esteem given him by others. In all four respects he has been tied to a model which he would scarcely accept since it would constitute a serious restriction on his options and opportunities. Clones do not differ from monozygotic twins in terms of their genetic duplicate but in the source of the duplicate. In the first case this source is an anonymous and responsibility-free entity, biological luck; in the second case it is an individual who has a name and who can be criticized for the fact that one is a genetic copy, but perhaps also for the fact that one is to a certain extent an unloved, perhaps even a false, copy.

As a clone one does not develop into a genetic twin in a chronologically parallel and consequently open manner but according to a model against which one might

---

[10] Kant (1790), § 40.

even be explicitly measured. One sacrifices much of one's own freedom for someone else's, although this other person has already had enough choices. As a result this constraint on freedom is very difficult to justify. In Habermas' slightly exaggerated formulation, the clone is a kind of slave because he 'can assign some of the responsibilities, which he would otherwise have to bear, to other people'.[11] In any case, the parent shoulders a responsibility which he most likely cannot bear and should not bear.

An interculturally recognized legal principle defended by Kant and reaffirmed in a celebratory manner by the United Nations is human dignity. Does cloning put it in question? There can be no doubt that a child produced through cloning possesses this dignity because human beings are entitled to it simply because they are human beings. Doubt cannot be directed towards the offspring but towards the way they are produced. In this respect the clone is instrumentalised in the interests of someone else. To avoid this instrumentalisation one has to be independent from the coercive arbitrariness of other people, something that can only be guaranteed by biological contingency.

Another counterargument, that genetic diversity is necessary for the survival of our species,[12] is not sound. There are around three to four monozygotic twins per 1,000, so that there are already today several million genetic duplicates. From the biological point of view, even thousands of clones would bring about an almost irrelevant increase of that number. It is not the collective, the species, which is to be protected. It is only the instance which eventually counts, the individual person.

By contrast, two other counterarguments are sound: on the one hand, cloned children constitute the first stage of designer children, thus they are not a first stage towards eugenics but already part of it. This is the case even if one distinguishes 'negative eugenics', eradicating hereditary diseases, from a positive one which selects desirable hereditary traits. In addition, one should not underestimate the value of stable family relations for the wellbeing of the children. Cloning would exacerbate some of the already existing risks associated with relationships: fathers could become the twins of their sons, grandparents the genetic parents of their grandchildren and mothers could give birth to a genetic twin of their own. Of course, some of today's risks are difficult to avoid; reproductive cloning, however, would increase them dramatically. Our societies are based on solidarity, of which they will, if necessary, bear the consequences; thus the legal-ethical conclusion is clear: in light of both the need to protect children and also to protect our society even a cloning that was of a low medical risk could only be permitted by an irresponsible legislator.

---

[11] Habermas (1998).

[12] Zimmer (1998).

14.4   Excursus: Gift, Not Property

One of the often quoted communitarians, the professor of government at Harvard University, Michael Sandel, reminds us that children are to be treated as a 'gift, not property'. He believes this aspect offers an alternative to the 'language' of fundamental and human rights which I prefer, and he goes on to claim that even now there are frequent and sundry violations of 'his' criterion of an appropriate relationship to children.[13] Though it is obviously right that children are no property two questions arise:

First we must ask whether each and every violation of Sandel's criterion, even the smallest one, is so legally relevant that first of all the legislator then, perhaps within the framework of legislation, the executive power and in cases of dispute the courts may and perhaps even should intervene. The answer will probably be similar to that to reproduction: the rearing of children is an intimate area in which the legal order may only intervene in the most extreme cases, for instance in the case of severe neglect or of physical or perhaps even psychological violence. If ambitious parents 'drill' their children too early and too hard, as criticized by Sandel, it should rather prompt relatives and friends or even public opinion to object rather than to encourage the legislator to introduce coercive measures.

Since scarcely any violation could be legally relevant, a discriminating criterion is needed. And here it is after all that 'language' of fundamental and human rights, including the fundamental rights of children, about which Sandel is sceptical, that is called for. It does not stop us from sharply criticizing this ownership mentality vis-à-vis children, even as social critique, a mentality that does not however require that a coercive legal order intervene. Legal ethics is not the same as overall social ethics, but simply constitutes an essential core area.

The second question is directed towards the justification of the ethics of 'gift, not property'. Again, the language of fundamental and human rights seems to be appropriate. After all, the ban on instrumentalising human dignity already includes a ban on using human beings, and by extension children, for the purposes of others.

14.5   'Therapeutic' or 'Research' Cloning?

Many objections disappear when it comes to the second type of cloning. However, the difficulties already begin with the semantics. Should one use the term 'therapeutic cloning' or rather 'research cloning' or prefer to speak of the 'deliberate proliferation' of a 'totipotent unit of cells'?

The expression 'therapeutic cloning' points to the promise of healing which research scientists themselves want to uphold. It also makes it easier to gain the approval of society, though this promise is not backed up by actual research.

---

[13] Sandel (2003).

It describes a far-off hope but by no means a present reality. Anyone who ignores this, will fall foul of a fourth type of mistake, the humanitaristic fallacy. This is because research focuses on (a) the preliminary work for (b) new opportunities of (c) a concrete therapy. The so-called therapeutic cloning is really the preliminary to the preliminaries for future assistance, a very paratory work.[14] For it is not so much fundamental research which can be expected to produce new insights but that applied research which is oriented towards medical assistance and which wishes to make diseases like childhood diabetes, Parkinson's disease and multiple sclerosis treatable by using embryonic stem cells.

The alternative expressions 'deliberate cell proliferation' and 'totipotent unit of cells', by contrast, uphold the argumentative sovereignty of biology. They sound like pure natural science without prompting those legal-moral questions which some people shy away from. After all, human embryos force limits on research according to the law actually in force, at least in many countries. And the legislation does so for good reason. One is not permitted to end someone else's human life – not even for research purposes or therapeutic purposes. In short, the expression 'therapeutic' cloning claims too much, 'deliberate cell proliferation' says too little. Therefore, it is recommended that we follow the example of the US Kass Commission and use the term 'cloning for biomedical research', perhaps also 'cloning for therapeutic research', abbreviated as 'research cloning'.

What distinguishes it from reproductive cloning? This leads on to the additional question: are there factors sufficiently exonerating to make this research acceptable in the legal and ethical perspective? Reproductive cloning pursues a private, not always worthy intention; research cloning, by contrast, pursues a thoroughly public and in the long term humanitarian, medical interest. It serves progress in the application of knowledge which has good prospects of helping doctors in their work in the long term.

However, the procedure itself is initially identical, which means one has to ask oneself whether the intention alters the action to such an extent that it changes its legal-moral significance: from a minus to a plus, or at least from minus to neutral. The current ethical and legal debate has not reached agreement on this. In the Kass Commission, too, competing positions prevent a joint stance.

One side, the utilitarian supporters of research cloning, argue in six steps; firstly, they consider that embryos in an early stage are worthy of protection but not to an unlimited degree. Secondly, they consider a utilitarian balancing of legally protected interests to be admissible. Thirdly, they dispute a destructive intention ('not created for destruction') and stress fourthly a high-ranking purpose, service to life and medicine even if, fifthly, they qualify this in a honest but almost masked manner as only being possible to a certain extent ('may come from it'). Sixthly, from this they draw the typically utilitarian conclusion that any possible major good has more weight than moral objections. The other position, which people like to call

---

[14] Cf. Höffe (2003), 44 ff.

deontological, believes that research cloning is a moral injustice as nascent human life would be exploited and destroyed. This would be reprehensible even in the case of so-called good intentions.[15]

For an ethical assessment of the controversy three questions are important. In contradiction to a widely held opinion all three questions are not truly interested in the dispute between utilitarianism and deontology which means that the popular political strategy becomes invalid: if the moral philosophers cannot agree here then people are free to support the deontologists refusal or the utilitarian support of research cloning. And the next step of the popular strategy says: as long as the legislator has no good reasons to forbid research cloning he has to permit it and to allow researchers to decide whether they will take part in research cloning or not. However, neither of the three relevant questions enters the dispute between utilitarian and deontological ethics.

When it comes to the first question, a distinction must be made between the significance of the purpose, between action-internal and action-external purposes. In the first case the purpose is a major or constitutive part of the action, in the second it is not. When I give someone money, the main issue is whether I am paying him for work he has done or whether I am trying to get him to violate his official duties; in the one case I am making a payment, in the other I am bribing someone. In both cases the intention lies in the definition of the action itself; it is an action-internal or action-constitutive intention without which the action remains under-determined. The situation is different in the case of the (other) intention in which I intend *to* bribe. It may be a good intention: to save somebody from an unjust punishment, or a bad one: to get an unwarranted advantage; in both cases it is action-external. This does not change the fact that, firstly, bribery has taken place and, secondly, that bribery is reprehensible from both the ethical and the legal point of view. Similarly, a conscious untruth, a lie, is reprehensible even if it is excusable for being used in an emergency situation ('white lie').

For the cloning debate it is easy to give an answer to the first question: both elements, reproduction and research, are not constitutive purposes of the action but rather external to the process of cloning. Even when a clone is produced for pure therapeutic research, it cannot be denied, in the case of a fertilized human being, that it constitutes the early stage of a human embryo. Consequently, the intention cannot change the moral significance but may call for forbearance in the case of truly therapeutic cloning. A humanitarian intention does not, therefore, ease the burden on research cloning, at least not to a sufficient degree.

The situation would be different if research cloning – the second question – were distinguishable, independent of its external intention, i.e. 'clone-internal', from reproductive cloning. Here the answer is not so easy because, while several points are not contentious, one is. It is not disputed that after completed fertilisation a living being is produced, i.e. a being that lives from itself, that organises and replicates

---

[15] The President's Council on Bioethics (2002).

itself in accordance with its own individual genome. Nor is there any dispute about the fact that it is not a sub-human being. Whether *in vitro* or *in vivo,* the developing being belongs to *our* species and to no other. Finally, there is agreement that within the framework of a complex development, implantation in the uterus is important since it affords food and protection, and probably emits activation signals even if, to my knowledge, this has not been conclusively proven. Therefore, the biological fact that even an in vivo egg cell is a living entity of the human species, is undisputed. What is disputed is its legal significance.

Concerns about an over-assessment of the significance of the further development of the egg cell already begin with the *biological* condition that reproductive and research cloning have the same intermediate product. That the product itself is available for implantation in a uterus, i.e. to being used for reproductive cloning, reinforces our assessment that the intention for cloning is not constitutive.

The second concern has to do with the *legal* side, the worthiness of protection which is characteristic for humans. It is important that this depends neither on a specific performance or property or on recognition by third parties. Human life is already deemed worthy of protection because it is *human.* In this respect research cloning seems to be even more troubling. After all, it voluntarily produces *human* embryos with the explicit goal of denying them, after some time, the possibility of further human development. This clearly instrumentalises human life for third parties, something which is very difficult to justify because our legal tradition forbids it and only permits it in exceptional cases where life stands against life, a balanced assessment of legally protected rights.

In any case, the third and decisive question is: What status does the intermediate product, which is open for both reproduction and research, enjoy?[16] This simple question confirms that the point of dispute lies outside the dispute between utilitarian and deontological ethics, and rather in the legal status of the human unit of cells. If it is human life from the very outset, then where does the utilitarian balancing of life worthy of protection actually begin?

First, a comment as to the relevance of this question is useful: Ethics is hereby brought into truly new territory, though unlike Hans Jonas' pathos[17] not into the emotion-laden territory that insists on fundamentally new moral principles. Rather, due only to the availability of new techniques, an old principle, one that has been acknowledged beyond cultural borders, must be put once again into question: Who exactly is this being who deserves a fully protected right to life (as well as the human dignity on which this protection is based)? It is not at all a rhetorical question since the boundaries of the group of beings, which humans comprise according to the definition of human rights, are no longer undisputedly clear-cut. This is not due to the fact that 'human being' is not a purely biological term, a label granted

---

[16] For the biological side cf. Nüsslein-Volhard (2003); for the legal and ethical side see Höffe et al. (2002); for a comprehensive debate cf. Damschen and Schönecker (2003).

[17] Jonas (1979).

by nature itself, but rather a culturally dependant ascription. It has been long acknowledged as a foundation of our legal culture that to belong to the human race is all that is required, that especially the poorest, the ill-treated, the emotionally or psychologically wounded and the socially despised, as well as the severely mentally disabled and the worst criminals are to be granted the right to life as well as human rights. It can also be shown by using medical imaging techniques that a developing embryo after a certain point bears a visible resemblance to a human being, though at first admittedly in a way that still requires interpretation.

Additionally, it is the case that yet to be born children can be named as heirs and are thus in effect members of a law-based community. Abortion is, according to German law, illegal though it goes unpunished, but only because of a normatively complex situation which biomedical research certainly cannot claim: The perpetrator (the mother) and the victim (the fetus) together constitute a double unity, which the sword of justice can but poorly protect, though severely wound.[18]

For religiously-minded people, a human being possesses inalienable rights because he is made in God's image. A religiously and world-view neutral, secular law-based society does not deny its Judeo-Christian background, but nevertheless does not build upon such arguments. It rather calls upon Kant, who bases human rights and human dignity on reason alone, more specifically on practical, not theoretical reason. He does not appeal to the rational but to the moral faculty. Even so, it remains disputed who exactly in this sense is to be considered human. Two fundamentally different views are at variance here.

According to the individualistic view, the only individual who qualifies as human or as a person, is one that possesses the ability to freely and reasonably determine itself, an ability which a young foetus obviously lacks. In the species view, on the contrary, humans are granted dignity as persons because normal members of the human family possess certain characteristics such as, for example, self-consciousness or self-respect. And precisely here lies a fundamental problem: May one follow the individualistic view, or must one follow the species view?

The species view is strengthened by the fact that the phrase 'human being' indicates an individual of a natural kind, whose members have abilities, which are denoted by the biological species name *Homo sapiens*, itself qualified by an additional *sapiens*. These members have the benefit of the faculties of speech, self-consciousness and reason. This fact must however be qualified in three respects: (1) that the members of the species have the faculties at their disposal is true for most of them but not for all, for instance not for some of the mentally challenged; the faculties are at their disposal (2) insofar as they have already gone through the required development, which is not the case for babies; and (3) inasmuch as they actively put these faculties to use, instead of letting them slumber, as when one is asleep. Concerning (2) and (3) the relevant persons have the capacities of speech, self-consciousness and reason. But even in case (1), concerning disabled beings,

---

[18] Cf. Höffe (2000).

we refer to them as members of the species and call them human beings, from the onset of life until death. Whoever is a member of this natural species, which as a species boasts of a (more than minimal) capacity to reason, is human. And because it is the *sapiens*, the gift of reason, that is the deciding factor, this is not a case of speciesism, as Peter Singer claims.[19] Another natural species which is capable of moral responsibility would be, as far as its members are concerned, placed on the same level as humans.

The above mentioned topical argument speaks against the individualistic ascription. Against a widely accepted understanding of the terms 'human' and 'human rights', the individualistic view denies to large sections of the population both the claim to be human and the associated claim to inalienable human rights. In order to avoid this all too obviously self-destructive consequence, the penal law theorist and philosopher Reinhold Merkel[20] has suggested the alternative principle of species solidarity. Two arguments however speak against it:

According to the first, reason-pragmatic argument, one does not venture into unknown territory with unproven instruments. The principle of species solidarity however is disputed and therefore normatively uncertain. Neither the concept of solidarity,[21] nor its possible link to the species, nor the extent of that link are themselves any less disputed.

The second argument is certainly the more important one. Because solidarity involves a clearly weaker obligation than does human dignity, too many beings are put in danger of losing their right to life to other people or goods, just as in the individualistic view. At least embryos and even those well along in their development as well as newborns and even the severely disabled lose their right to an uncompromised protection of their lives.

So, once again, when exactly does human life begin? Merkel, who has criticized protecting life at a seemingly too early stage, concedes that the process of human development is a continuous one. He claims though, that we can make distinctions that are as little arbitrary as when one distinguishes between a 'short' man who is 1.50 m tall and a 'huge' man who is 2.50 m tall, or between a 'pitch black' night and a 'brightly sunny' day. This objection supposes however that the development of a fertilized egg is in a crucial respect quantitative (dark-light). In reality, it is a qualitative development, coming from and out of itself.

No one denies that a complete life-program is not yet a fully developed human being. This program however contains something fundamentally different than the mere potential of, say, a block of marble stone intended for a statue. No statue carves itself out of its block of stone. Without the artist, it remains a simple block of marble, period. At best, a contingent natural erosion will give it a certain, but also random and non-programmed shape, due to which perhaps a given rock formation might resemble a plant, another an animal or a human being. Fertilized eggs,

---

[19] Singer (1979), Chapter 3.

[20] Merkel (2001).

[21] For a first clarification of the concept cf. Höffe 2007a, Chapter 3.6.

on the contrary, develop from within; only they, not the sperm or the egg themselves, go through an auto-determined life process. We may say, sperm and egg separated have a passive potentiality, the fusion of both an active potentiality. Owing to their auto-determination, since the fusion of sperm and egg, the new entity is in an elementary (though not optimal!) way free: it is not subject to foreign laws or programs, but only to their own law and to their own program.[22]

If it is granted that the development of a human being is fully programmed as soon as sperm and egg fuse, then this is a good, though not irrefutably compelling, argument for already speaking of a human being in the strong and protective legal sense. The question is not simply rhetorical but really open: does the scientist, who through negligence lets an *in vitro* embryo die, commit involuntary manslaughter? Even so, it can no longer be doubted that in the strong sense of the term a human life is indeed at issue here. To go about in a negligent manner is at least to carelessly endanger human life and possibly to let it die due to negligence.

One can certainly ask oneself who one should rather save out of a burning laboratory: the *in vitro* engendered, living embryos or an unconscious infant. That the infant will plausibly be given priority does not however allow any conclusions regarding the main question of whether embryos, insofar as they are human, may be killed; this is because the duty of assistance is less binding than the prohibition of killing. Analogous to the (disputed) case of the permissibility of white lies, it is never permitted to kill someone in order to help someone else. Moreover, as mentioned above, one must be wary of committing the humanitaristic fallacy: Research that requires the destruction of embryos is (a) preliminary work aiming at (b) finding new alternatives for (c) a possible cure. It is but the preliminary to preliminaries to a future solution.

To repeat the decisive point: the supporters of research cloning grant a value worthy of protection, but only a relative one. The other side contradicts this by arguing that what some people dismiss as a 'mere cluster of cells' carries in itself from the very outset, as a fertilized egg cell with a double set of chromosomes will, the full life programme for the development of a human being. It is true that, in contrast to a bulb or to the eggs laid by fish, frogs and birds, mammals and human beings are dependent on a sophisticated environment, the womb. But the life programme is clearly a human one; it is equally clear that the human programme has already begun its human development. It requires no additional improving change, no fine tuning. In this respect, it takes place in a continuum: because of self-control but not according to external laws or programmes but according to its own laws and programmes. Therefore, one can hardly avoid saying that the embryo is not a potential but an actual human being.

Admittedly it is a further question to decide whether this human being is a potential person who will become a person at some future stage or whether it is a

---

[22] For an introduction to and overview of the rarely discussed Muslim side of the debate: Ilkilic, 2004.

person with potentialities that means it is already a person but cannot exercise properties as a person until (much) later. But is this question decisive? One could prefer the concept of capacity to that of potentiality and add that all human beings have at any stage after fertilization the capacity for personhood, even the foetus and the mentally challenged. But the alternative concept has a similar problem: A foetus and a mentally challenged person has the capacity in a very different way than the 'ordinary persons'. Thus, it is necessary to make distinctions and a sceptic might question whether capacity to personhood is a sufficient argument for a full-fledged protection of that entity.

However, since that relativisation, which the opponents of identifying fertilized eggs as human beings justify by referring to the purpose of the womb, only takes place within the framework of a basic continuity, research cloning remains – to put it cautiously – ethically and legally worrying.

All three authorities, our legal order, medical ethics, starting with the Hippocratic oath, and moral philosophy distinguish between the morals of law, which people are bound to uphold for each other, and voluntary additional contributions of the morals of (meritorious) virtue. They add that in the name of the duty to help (based on the morals of virtue) no human life may be killed. Except in the case of self-defence, one may not even harm human life.

Fortunately, opportunities for reprogramming have appeared recently which avoid using the human embryo and, therefore, calm the corresponding concerns.[23] Because of this and because the road from embryonic stem cells to effective therapies is still a very long one, and because even experts are sceptical about its real therapeutic potential, people should move away from their fixation on previous procedures and release their scientific creativity, the curiosity mentioned at the beginning, for methods which are uncontroversially acceptable from both the ethical and the legal point of view.

References

Aristotle (1960). *Topics*. E. S. Forster (ed.), London/Cambridge, MA: Harvard University Press.
Burley, J. (ed.) (1999). *The Genetic Revolution and Human Rights. The Oxford Amnesty Lectures 1998*, Oxford: Oxford University Press.
Damschen, G. and D. Schönecker (eds.) (2003). *Der moralische Status menschlicher Embryonen. Pro und contra Spezies-, Kontinuums-, Indentitäts- und Potentialitätsargument*, Berlin/ New York: Walter de Gruyter.
Dworkin, R. (1996). *Freedom's Law*, Oxford: Oxford University Press.
Fletcher, J. (1972). *The Ethics of Genetic Control. Ending Reproductive Roulette*, New York: Anchor Books.
Forum Diderot (1999). *Faut-il vraiment cloner l'homme?*, Paris: PUF.

---

[23] Cf. Rapp (2003).

German Reference Centre for Ethics in the Life Sciences (DRZE) (ed.) (2003). *Klonen in biomedizinischer Forschung und Reproduktion. Wissenschaftliche Aspekte – ethische, rechtliche und gesellschaftliche Grenzen* (Reader), Bonn, Germany.

Geyer, Chr (2001). *Biopolitik. Die Positionen*, Frankfurt/M, Germany: Suhrkamp.

Habermas, J. (1998). "Sklavenherrschaft der Gene. Moralische Grenzen des Fortschritts", *Süddeutsche Zeitung*, January 17, 18, p. 13.

Höffe, O. (2000). *Moral als Preis der Moderne. Ein Versuch über Wissenschaft, Technik und Umwelt*, 4th edn., Frankfurt/M, Germany: Suhrkamp.

Höffe, O. (2003). *Medizin ohne Ethik?*, 2nd edn., Frankfurt/M, Germany: Suhrkamp.

Höffe, O. (2007a). *Democracy in the Age of Globalization*, Dordrecht; German: *Demokratie im Zeitalter der Globalisierung*, 2nd edn., Munich, Germany: C. H. Beck Verlag, 2002.

Höffe, O. (ed.) (2007b). *Lexikon der Ethik*, 7th edn., Munich, Germany: C.H. Beck.

Höffe, O. et al. (2002). *Gentechnik und Menschenwürde. An den Grenzen von Ethik und Recht*, Cologne, Germany: DUMONT Literatur und Kunst Verlag.

Honnefelder, L. and D. Lanzerath (eds.) (2003). *Cloning in Biomedical Research and Reproduction. Scientific Aspects-Ethical, Legal and Social Limits*, Bonn, Germany: Bonn University Press.

Ilkilic, I. (2004). "Der moralische status des Embryos im Islam und die wertplurale Gesellschaft", in: Baumann, E. (ed.), *Weltanschauliche Offenheit in der Bioethik*, Berlin: Duncker & Humblot, pp. 162–176 (with attendant bibliography).

Jaenisch, R. and I. Wilmut (2001). "Don't Clone Humans!", *Science* 291, March 30, 2552.

Jonas, H. (1972). "Biological Engineering – A Preview", in: Jonas, H. (ed.), *Philosophical Essays. From Ancient Creed to Technological Man*, Englewood Cliffs, NJ: Prentice-Hall, pp. 141–167.

Jonas, H. (1979). *Das Prinzip Verantwortung. Versuch einer Ethik für die technologische Zivilistion*, Frankfurt/M, Germany: Suhrkamp. English: *The Imperative of Responsibility: In Search of Ethics for the Technological Age*, Chicago, IL: University of Chicago Press, 1984.

Kant, I. [1902 ff.] (1790). *Kritik der Urteilskraft*, *Gesamelte Schriften*, vol. V, Berlin: Königlich Preußischen Akademie der Wissenschaften, pp. 165–485.

McLaren, A. (ed.) (2002). *Cloning*, Strasbourg, France: Council of Europe Publishing.

Merkel, R. (2001). "Rechte für Embryonen? Die Menschenwürde läßt sich nicht allein auf die biologische Zugehörigkeit zur Menschheit gründen", in: Geyer, Chr. (ed.), *Biopolitik. Die Positionen*, Frankfurt/M, Germany: Suhrkamp, pp. 51–64.

Nüsslein-Volhard, C. (2003). *Wann ist der Mensch ein Mensch? Embryologie und Gentechnik im 19. und 20. Jahrhundert*, Heidelberg, Germany: C. F. Müller Verlag.

The President's Council on Bioethics (2002). *Human Cloning and Human Dignity. An Ethical Inquiry*, Washington, DC.

Rapp, U. (2003). "Alternatives to Therapeutic cloning: Evolution of Pluripotency", in: Honnefelder, L. and D. Lanzerath (eds.), *Cloning in Biomedical Research and Reproduction. Scientific Aspects-Ethical, Legal and Social Limits*, Bonn, Germany: Bonn University Press, pp. 421–425.

Rawls, J. (1993). *Political Liberalism*, New York: Columbia University Press.

Sandel, M. (2003). "The Ethical Implications of Human Cloning", in: Honnefelder, L. and D. Lanzerath (eds.), *Cloning in Biomedical Research and Reproduction. Scientific Aspects-Ethical, Legal and Social Limits*, Bonn, Germany: Bonn University Press, pp. 465–470.

Singer, P. (1979). *Practical Ethics*, Cambridge: Cambridge University Press.

Zimmer, D. E. (1998). "Eineiige Zwillinge sollen Zufall bleiben", *Die Zeit*, January 12, 28.

Part IV
Theological Perspectives

Chapter 15
Stem Cells from Human Embryos for Research? The Theological Discussion Within Christianity

Lars Østnor

Abstract The author presents a survey of what churches and some theologians from different denominations are thinking with regard to the use of embryonic stem cells for research. He finds that there is an agreement among Orthodox churches, the Roman Catholic Church and several Protestant churches raising objections against such research. However, some Protestant churches in Europe and USA support it. The most used argument against is the specificity argument: Humans are unique creatures, brought into existence by divine activity and with the right to life and to not be harmed. Other arguments are the potentiality argument (embryos have the potentiality of becoming developed human beings) and the argument pointing to the continuity in development from fertilization and onwards. Churches supporting embryonic stem cell research refer to a graduality in the development of an embryo and to the new possibilities for treatment of serious diseases by using embryonic stem cells.

Keywords Church, embryonic stem cells, human being, human embryo, stem cell research

15.1 Introduction

We are about to investigate a new and old problem on the agenda of Christian ethics. Since the publication in November 1998 of the isolation of human pluripotent stem cells from embryos, there has been a comprehensive debate also among representatives from different denominations and churches concerning the moral status of embryos and

MF Norwegian School of Theology, Box 5144 Majorstua, 0302 Oslo, Norway.
e-mail: Lars.Ostnor@mf.no

L. Østnor (ed.), *Stem Cells, Human Embryos and Ethics: Interdisciplinary Perspectives.*
© Springer Science + Business Media B.V. 2008

the possible use of them as sources for stem cells. To some extent, however, it might be said that this is an old question become new. The question of the responsibility towards human embryos has been thoroughly discussed among theologians and church people for several years with regard to abortion, in vitro fertilization and research on fertilized eggs. The new aspects of this burning question are the possibilities within scientific bio-medical research and clinical therapy which might result from the use of human stem cells from embryos. The potential utility of research on human embryonic stem cells has sharpened the ethical debates and divided citizens, politicians and professional people in societies around the world. It might therefore be asked: Is it a splitting question also among Christians, or are the church bodies united in a common position?

The main question in the contributions from the churches – and which I want to clarify in this chapter – is this: Is research on human embryonic stem cells for medical purposes ethical justifiable or not? And what is the ethical argumentation for a support or for a refusal of such research? If research for medical reasons is seen as ethically acceptable, what is more precisely meant by acceptable research?

In my contribution I will not distinguish between fertilized egg, zygote, blastocyst, preembryo and embryo as is sometime done in different publications. Some authors use such conceptual differentiation when they are ethically interested in indicating that there is a biological development with corresponding variation in moral status. In this context I prefer to make use of a common concept for the relevant period of early human life. I therefore speak of an embryo, understood as a human body from fertilization to the end of the 8th week.

In general I consider statements from church authorities to be more representative for the denominations than statements from single individuals. Consequently, I give priority to documents from the relevant church institutions and supplement them with contributions from some persons within the respective traditions.

For obvious reasons the material which will be analysed is mainly from the period after 1998. It will consist of church declarations, statements and resolutions combined with books and articles from theologians.

The engagement from the side of churches and theologians has been strong and productive, especially in the USA and some European countries. The positions and the reflections can therefore be looked upon only by focusing on some central and substantial contributions.

My purpose and corresponding method is here mainly analytic-descriptive and not normative.

15.2   The Orthodox Churches

Interesting sources into an Orthodox way of thinking are two statements from the Holy Synod of Bishops of the Orthodox Church in America. The first one from 2001 deals with *Embryonic Stem Cell Research* and the second from 2002 with *The Cloning of Human Embryos*.[1] In the first case the bishops begin their comments

---

[1] The Holy Synod of Bishops October 17, 2001 and the Holy Synod of Bishops January, 2002.

to the public debate over research on embryonic stem cells with a clear and distinct utterance:

> From the perspective of Orthodox Christianity, human life begins at conception (meaning fertilization with creation of the single-cell zygote). This conviction is grounded in the Biblical witness (e.g., Ps 139:13–16; Isaiah 49:1 ff; Luke 1:41,44), as well as in the scientifically established fact that from conception there exists genetic uniqueness and cellular differentiation that, if the conceptus is allowed to develop normally, will produce a live human being. Human life is sacred from its very beginning, since from conception it is ensouled existence. As such, it is "personal" existence, created in the image of God and endowed with a sanctity that destines it for eternal life.

The members of the synod published their ethical position on the background of President Bush's decision that would allow scientific research on some 60 lines of existing stem cells from human embryos which were destroyed after the harvesting of the cells. The synod evaluates this decision as a compromise between on the one hand protection of human life at every stage of its development and on the other hand studies of the potential therapeutic benefits from pluripotent stem cells. Still the synod maintains that the destroying of those embryos is an evil act and that 'we may not profit from evil even to achieve a good and noble end'. The members fear that the result will be 'ever more utilitarian manipulation of human life' and 'an increasingly slippery slope'.

The bishops oppose embryonic stem cell research with political, medical and theological arguments in the following way: First, embryo research is supported by pro-abortion activists wishing preservation of so-called 'abortion rights'. A governmental rejection of funding such research would indicate that human life starts at conception and that an embryo is not a clump of tissue. Second, strong pressures to promote embryonic stem cell research by means of law and money come from biotech and pharmaceutical industries and market forces will determine the access to and the costs of medicine and therapy based on stem cells. Third, experience from the past shows that claims from scientists that use of fetal cells will give medical treatment of serious illnesses, have failed. Such claims regarding stem cells may also be exaggerated. Fourth, the embryonic stem cell research is to be seen as a point on a utilitarian slippery slope leading to a tragic devaluation of human life, if the trend is not reversed. Fifth, it has been a universal principle accepted by scientists that 'no experimentation should be undertaken on human subjects without the subjects' informed consent'. But such a consent can not be given by an embryo, nor has the mother or anyone else the right to grant it. Sixth, cloning of human embryos is necessary in order to produce sufficient copies of cells for research. Cloning of animals has shown serious genetic defects and human cloning has the same danger and should be permanently banned. Seventh, there have been published reports from researchers, finding that adult stem cells and cells from placentas and umbilical cords are as much if not more promising than embryonic stem cells.

Central elements of the conclusion are these:

[…] we firmly reject any and all manipulation of human embryos for research purposes as inherently immoral and a fundamental violation of human life. We call upon the President and the Congress of the United States to restore and maintain a total ban on ESCR (embry-onic stem cell research). Furthermore, we encourage the scientific community … to devote their energies and resources to discovering, harvesting and utilizing non-embryonic stem cells, including those derived from adults, placentas and umbilical cords.

Above all, we urge our faithful, together with the medical community and political leaders, to return to the spirit of the Hippocratic Oath: primum non nocere, 'First of all, do no harm.' Embryonic stem cell research results in unmitigated harm. It should be unequivo-cally rejected in the interests of preserving both the sacredness and the dignity of the human person.

In the document from 2002 the Holy Synod of Bishops repeats and deepens its opposition against the cloning of human embryos, including both therapeutic and reproductive cloning. They maintain that all cloning is reproductive in the sense that it creates an embryo who is to be recognized as an individual human being. The members of the synod identify the crucial problem of the debate as the following: 'Is the pre-implantation embryo a unique human being?' And they try to present an argumentation for their own positive answer by combining bio-medical and theo-logical elements:

[…] even at the earliest stage of growth the cluster of blastomeres constitutes an entity characterized by both genetic and developmental individuality […] Although cellular dif-ferentiation is clearly observable from only about the eighth day after fertilization, from the very beginning each cell functions as part of an organic whole. And although the early embryo is a 'potential plurality' (i.e. it can twin), it is an actual organic unity, a specific and unique human individual, endowed with human nature and bearing the image of God.

Any living organism is characterized by growth and change. Thus it possesses both actual and potential aspects. The fact that much in the life of an embryo is potential does not alter its nature as a human being. Because the DNA or genetic code is fully present from fertilization (or, we must specify today, from the onset of embryonic growth), unique and individually differentiated human life is fully present, even though it has not been expressed as specific organs or capacities, and even though it may twin to produce multiple offspring.

The bishops conclude by urging 'a basic Christian truth': 'that from conception human life is a sacred gift, one called at every stage of its existence to grow toward eternal participation in the life of God'.

Several substantial elements are central in the Orthodox argumentation and sus-taining of the ethical position: (1) A creation theology according to which a human being is unique because of having been created in the image of God. (2) Conception as constitutive for the human individual. (3) The sacredness and specific dignity of human, biological life. (4) The human embryo has all its character present although much of it is only potential. (5) Good purposes can not be ethically justified when depending on use of evil means. (6) The meaning and aim of human life is 'the life of God' (theosis).

In their first statement the Orthodox bishops had referred to the theologian John Breck's book *The Sacred Gift of Life* (1998). For obvious reasons he does not explicitly examine the stem cell field. But he presents the fundamental values for a *theological* bioethics: the sacred character of human life, the sacrificial love of God and the call to holiness and deification (pp. 243 ff.). He marks a front against the distinction between an embryo and a so-called pre-embryo, the latter being a term for the period until the development of the 'primitive streak'. According to him the basic problem is the definition of 'individuation' and his own answer is that a pre-implanted embryo is an individuated, personal human being from fertilization. This is due to 'the undeniable evidence that the human "soul", the divinely bestowed dynamic of animation, is present from the very beginning' (pp. 139–143). A distinction between 'genetic' and 'developmental individuality' is impossible (p. 138). The zygote or embryo is body and soul from day 1.

In 2001 Demetri Demopulos applies the Orthodox bioethical principles on our topic more directly. He is aware of the great medical possibilities of research on embryonic stem cells, but he opposes therapeutic cloning because of the destruction of the embryos. Instead he recommends use of adult stem cells which according to him at that time seem to provide potential for the treatment of human diseases.

15.3 The Roman Catholic Church

Does the Roman Catholic Church present a corresponding, restrictive ethical evaluation as their Orthodox colleagues?

A central document for information about the official position of the Roman Catholic Church is to be found in *Declaration on the Production and the Scientific and Therapeutic Use of Human Embryonic Stem Cells* from Pontifical Academy for Life in August 2000. After having given a brief presentation of the recent scientific data on stem cells and the biotechnological data on their production and use, the text turns to three key ethical problems in this context. The first one, which is characterized as fundamental, is formulated in this way: 'Is it morally licit[2] to produce and/or use living human embryos for the preparation of ES cells?' The question is given a negative answer with the following justification:

(a) On the basis of a complete biological analysis, the living human embryo is – from the moment of the union of the gametes – a *human subject* with a well defined identity, which from that point begins its own *coordinated, continuous and gradual development*, such that at no later stage can it be considered as a simple mass of cells.

In this way, the starting point for the argumentative path is a reference to natural law, where 'natural' in this case is understood as a biological truth regarding the very beginning of human life. However, it is not quite clear what in this context is meant by 'human subject'. From the textual context it seems that it is used identically as the concept 'human individual'.

---

[2] Within Roman Catholic moral philosophy the concept 'licit' is used with the substantial meaning that something is ethical acceptable or permissible.

(b) Being such a 'human individual,' the embryo 'has the *right* to its own life; and therefore every intervention which is not in favour of the embryo is an act which violates that right'. The right to life and not to be harmed is accordingly ascribed to the embryo qua human being.

(c) The ethical evaluation is expressed by stating: 'Therefore, the ablation of the inner cell mass (ICM) of the blastocyst, which critically and irremediably damages the human embryo, curtailing its development, is a *gravely immoral* act and consequently is *gravely illicit*.' Destruction of embryonic life is in sharp contrast with a normative element within an ethics of rights.

(d) The ethical reasoning of the Roman Catholic Church also includes a reflection of the teleological aims compared to deontological duties in this case: *'No end believed to be good,* such as the use of stem cells for the preparation of other differentiated cells to be used in what look to be promising therapeutic procedures, *can justify an intervention of this kind.* A good end does not make right an action which in itself is wrong.'

(e) The last argument is that this position is explicitly confirmed in the Encyclical Letter *Evangelium Vitae* from Pope John Paul II in 1995, including a reference to the Instruction *Donum Vitae* published by the Congregation for the Doctrine of the Faith, which affirms:

> The Church has always taught and continues to teach that the result of human procreation, from the first moment of its existence, must be guaranteed that unconditional respect which is morally due to the human being in his or her totality and unity in body and spirit: 'The human being is to be respected and treated as a person from the moment of conception; and therefore from that same moment his rights as a person must be recognized, among which in the first place is the inviolable right of every innocent human being to life.

The second ethical problem dealt with in the declaration is this: *'Is it morally licit to engage in so-called "therapeutic cloning"* by producing cloned human embryos and then destroying them in order to produce ES cells?' Once again the answer is negative, because such praxis implies the same ethical problem as discussed above.

The third and final ethical problem is formulated in this way: *'Is it morally licit to use ES cells, and the differentiated cells obtained from them, which are supplied by other researchers or are commercially obtainable?'* Still the answer is negative, because 'the case in question entails a proximate material cooperation in the production and manipulation of human embryos on the part of those producing or supplying them'.

The declaration ends by proposing the use of adult stem cells. They represent 'a more reasonable and human method for making correct and sound progress' in research and therapeutic applications.

Already 13 years before this statement the Roman Catholic Church had published the already mentioned Instruction *Donum Vitae*.[3] This document contains some premises for the ethical position regarding the use of human embryonic stem cells.

---

[3] Congregation for the Doctrine of the Faith 1987.

A reflection on the nature and identity of the human embryo contains statements about a fertilized egg as for example these: 'It would never be made human if it were not human already' and 'how could a human individual not be a human person?' The conclusion with regard to respect for human embryos is this: 'since the embryo must be treated as a person, it must also be defended in its integrity, tended and cared for, to the extent possible, in the same way as any other human being as far as medical assistance is concerned'. There is no doubt that the Instruction claims respect, right to life and physical integrity for the embryo 'just like any other human person'. This evaluation is also applied to the question of the use for research purposes of embryos obtained by fertilization 'in vitro'. It is condemned as immoral (i) to produce human embryos destined to be only so-called 'biological material', (ii) to destruct such embryos obtained 'in vitro' for the sole purpose of research and (iii) to use them for risky experimentation. 'Every human being is to be respected for himself, and cannot be reduced in worth to a pure and simple instrument for the advantage of others.'

This doctrinal teaching of the Magisterium of the Roman Catholic Church has been disputed by theologians within this church. An example of this kind is to be found in the American theologian Margaret A. Farley (2001). She argues pro embryonic stem cell research. Her justification of this position is short, but it seems to consist of several elements: On the basis of recent embryologic studies she maintains that a human embryo at its earliest stages (before the development of the primitive streak or the implantation) can not be considered as 'an individualized human entity with the settled inherent potential to become a human being' (p. 115). On the same empirical basis she argues that fertilization itself is a process and not a moment, and that early embryos (including the time for harvesting stem cells from the inner cell mass) are not sufficiently developed and individualized in order to have personhood. But using them for certain research aims presupposes a sharp barrier between therapeutic cloning and reproductive cloning.

Other Roman Catholic theologians oppose this alternative of thinking by referring to two arguments (according to Carlberg 2004:140 f.): First, individualization is not a premise for protection of human embryos who deserve protection because of their potentiality. Second, individuality of a zygote can be scientifically based on the process of coordination between the cells.

15.4   The Protestant Churches

The Protestant contributions to our topic represent a broad spectrum of documents, delivered at different times and in various contexts. A great number of the texts have been published in Germany. They must be understood on the background of the country's restrictive legislation from 1990, which prohibits the destruction of embryos in connection with research. An embryo is here understood as 'the fertilized and for development capable human eggcell from the point of fusion between the nucleus of the gametes' (see Carlberg 2004:141). In May 2001 The German

Research Foundation (Die Deutsche Forschungsgemeinschaft) recommended that German scientists should be allowed to import pluripotent, embryonic stem cell lines from abroad, and if necessary, harvest stem cells from surplus embryos themselves. This proposal brought about a profiled statement from the Council of the Evangelical Church in Germany with the title *The protection of human embryos must not be reduced* (epd-Dokumentation 2001:1 f.). The Council defends the existing law and understands production of human, embryonic stem cells for research and the use of spare embryos for scientific purposes as inconsistent with the legislation. The declaration comments on and opposes five general arguments that are often used in the debate now proceeding:

(a) An important motivation for the research discussed is the possibility of developing new medical therapies. From a Christian point of view this aim is ethically legitimized (the commandment of neighbourly love). But a purpose with high priority shall not be realized for any price. The means which are used, must also be ethically acceptable. And all the promises about therapy ought to be seriously examined.

(b) The German Research Foundation has seen the introduction of in vitro fertilization as a fundamental decision and has maintained that there is no return to the right to life of embryos before this occasion. The Council supports the understanding of this introduction as a decisive new track and advises against it. But technical developments may be changed and political and ethical powers may lead to the end of a started process because of better insights.

(c) The German Research Foundation maintains that it is necessary to find a balance between a judicial protection of the embryo's life on the one hand and a judicial protection of the freedom of science on the other hand. However, the Council rejects such a balance, because the protection of the dignity and of the life of a human being, guaranteed in the legislation, belongs to the human embryo. Freedom of research has a limit, marked by the protected dignity of human beings.

(d) In the debate it has been maintained that it would be equivocally and ethically doubtful to let research be carried out abroad and at the same time claim the utility of this technology in Germany. The Council claims: The No to research on embryonic stem cells becomes credible when it is combined with a defence of the standards of the strict German legislation of embryo protection.

(e) To an unlimited protection of human embryos it is critically objected that it is in contrast to the legal regulation and the praxis of abortion. The Council admits that the protection of the embryo in vitro and in vivo is ethically linked together. But one needs to distinguish between an unforeseen conflict situation and a foreseen alternative of action in the laboratory. Concerning abortion cases there is no principal reduction of the protection of the unborn child.

In January 2002 the German Parliament (der Bundestag) decided to allow the import of embryonic stem cell lines from abroad (Carlberg 2004:141). Right after this decision the chairmen of the Council of the Evangelical Church and of the German, Roman Catholic conference of bishops commonly expressed their disappointment (Kock and Lehmann 2002). The approval would make it possible in the country to experiment with human stem cells which depends on the killing of embryos. The right to life and unlimited life protection of human beings from the point of fertilization is not anymore guaranteed. The decision is against the spirit of the legislation with its protection of embryos.

The social-ethical argumentation of the Council of the Evangelical Church in Germany is on the one hand typically contextual in the concrete, public debate within this society; on the other hand it presents substantial aspects with relevance to the discussion which had been going on and is still alive in several countries. Of current interest is especially its understanding of the relation between the ethics of the church and the legislation of a society.

In a document from 2002 The Advisory Commission for Public Responsibility of the Evangelical Church in Germany published *Argumentationshilfe* (Help for argumentation) in medical-ethical and bioethical questions.[4] Here one argues for an understanding of human beings as persons, based on God's unreserved acknowledgment, not on qualities and capabilities. Evangelical theology emphasizes the relational character of a person – with relation primarily to God, subsequently to fellow human beings and to oneself. If unborn life may be personal life, it is according to this a question of whether it has such relations (pp. 17–19). A human embryo is commonly seen as 'a developing human being', without any fixed point of time regarding the individual existence of a human. Consequently, the members differ when confronting the problem when and why an embryo must be defined as a human being: from the fertilization with the given, inherent process of development or when external conditions create possibilities for development (pp. 20–24). There is also a corresponding disagreement when dealing with the question of research use of surplus embryos after in vitro-fertilization, but a consensus in refusing the production of embryos in vitro for such purposes because of a one-sided, biological point of view. Reproductive cloning is unanimously rejected, but so-called therapeutic cloning is differently evaluated along the same lines as already mentioned. One part argues against therapeutic cloning by referring to the use of embryos as means for research purposes, which is contrary to human dignity. The other part accepts such cloning for medical reasons and sees human dignity as irrelevant because no human being is produced (pp. 30–33).

This document reflects a certain division within the Evangelical Church in Germany in large with regard to our topic. And it demonstrates that the dividing question is whether an early embryo is a human being or not and the criteria for answering this question in one way or another.

In France the federation of Protestant churches in 2003 rejected both reproductive and therapeutic cloning, the latter leading to a utilitarian view of human life (Carlberg 2004:144 f.). With regard to research on surplus embryos the federation avoids decision. On one hand such science may express a 'therapeutic solidarity'

---

[4] *Im Geist der Liebe mit dem Leben umgehen.*

corresponding to the Christian commandment of neighbourly love; on the other hand it may imply using human life as an instrument, being in conflict with the respect towards such life. The executive committee of the Church of Sweden gave in 2003 their support to the proposal from the government accepting the use of surplus embryos for stem cell research (Carlberg 2004:143 f.). Nor do they categorically reject therapeutic cloning. Scientific research is legitimized as participation in God's ongoing creation, but human dignity must be respected. Because it is not evident when such dignity is a reality, destruction of human life at an early stage may be justified.

The public discussion among Protestant theologians covers a broad spectrum of contributions. Only a limited number of them can be mentioned here. A group of nine evangelical theologians from Germany, Austria and Switzerland have expressed their common position by referring on the one hand to the freedom and plurality regarding ethical standpoints as a characteristic feature of Protestantism, on the other hand to the necessity of finding a compromise between the different opinions concerning research on human embryos.[5] They find three positions in the current debate: First, the demand for an absolute protection of embryos, justified by the reference to potentiality and continuity. Second, the demand for a gradual protection of embryos, in principle from the beginning of life, unlimited from the point of implantation. Justification of this alternative is the reference to external conditions for developing to an existing human. Thirdly, some people plead for an unlimited possibility of research on embryos, justifying the position by denying that early human life is a person and a human being with absolute protection. The divergence is identified as disagreement mainly with regard to the point of time when a developing human life is to be defined as a human being. The group itself supports the second line. According to this, the question of point of time can not be solved by biological knowledge, because such empiric reasons can only produce plausible indicators. Empirical and ethical aspects have to be combined. Decisive for these theologians is the fact that the ideas of human's image of God and human dignity, according to their view, can not be identified with a certain stage of the biological development. Both anthropological concepts must be attributed to human beings transempirically, but they express only an aspect of human existence and function. The image of God is in evangelical theology interpreted to be realized in the future and human dignity must include the element of the tasks of humans. To be a human being is consequently both 'Genom' (genome) and 'Geschichte' (biographical history). The other perspective of the argumentation is the positive evaluation of scientific research and its therapeutic cure of diseases, turning against an unrealistic effort of legitimation. Evangelical theology opens up for individual solutions of ethical conflicts in specific situations. The proposal of the group for a way out of

---

[5] Anselm, R. et al. 2002. The members of the group were the Professors Reiner Anselm, Göttingen, Johannes Fischer, Zürich, Christofer Frey, Bochum, Ulrich Körtner, Wien, Hartmut Kress, Bonn, Trutz Rendtorff, München, Dietrich Rössler, Tübingen, Christian Schwarke, Dresden and Klaus Tanner, Halle-Wittenberg.

the existing dilemma is an acceptance of research on surplus ('orphan') embryos because of their missing possibility for development. In the same way they evaluate the use of already existing stem cell lines. However, they refuse the production of embryos for research aims.

Several Protestant churches in the USA have commented upon the use of human embryonic stem cells in research and therapy. Southern Baptist Convention in June 1999 strongly opposed the proposal from The National Bioethics Advisory Commission that the ban on public funding of human embryo research should be removed (Waters and Cole-Turner 2003:179 f.). Such an effort expresses a utilitarian ethic, willing to 'sacrifice the lives of the few for the benefits of the many'. The prohibition is, according to the resolution, built upon legal and ethical norms against misusing human beings for research purposes, including principles in the Nuremberg Code, the Helsinki Declaration and the United Nations Declaration of Human Rights. A premise for the argumentation is that 'protectable human life begins at fertilization' and that human embryos are 'the most vulnerable members of the human community'. The resolution calls for support for alternative research and treatments without the destruction of human embryos.

In 2001 the General Assembly of the Presbyterian Church (USA) resolved a statement dealing with stem cell and fetal tissue research.[6] The moral status of human embryos is defined by maintaining that they have the potential of personhood and therefore deserve respect. Attention to this respect requires that the goals aimed at by using embryos can not be reached by other means. But such respect shall not have priority above the medical help towards persons with pain and disease. Use of embryos, however, should be limited to surplus embryos. The conclusion states: 'With careful regulation, we affirm the use of human stem cell tissue for research that may result in the restoring of health to those suffering from serious illness'. The potential lifesaving and health recovering is evaluated to be of greater value than the life of 'embryos that do not have a chance of growing into personhood'.

The General Synod of the United Church of Christ resolved in 2001 a resolution with the programmatic heading 'Support for Federally Funded Research on Embryonic Stem Cells' (Waters and Cole-Turner 2003:181–184). It contains no contra-arguments against such research. The main pro-argumentation is that 'Such research may enable the development of new approaches to diagnosis, prevention, and treatment of some of our most devastating diseases such as Parkinson's, Alzheimer's, Juvenile Diabetes and heart disease'. The theological argument pro is a reference to Jesus as an example for the support of healing and caring for the sick and disabled.

United Methodist Church the same year supported the proposal of banning all forms of human cloning.[7] The general argument for this position is that the creation

---

[6] *Statement on the Ethical and Moral Implications of Stem Cell and Fetal Tissue Research.*

[7] The document *Urgent Action Alert: Urge Senators to Support Complete Ban on Human Cloning*, see Waters and Cole-Turner (2003:177 f.).

of 'human life' for exploitation and destruction is 'displaying a profound disrespect for life'. A theological argument is only scarcely touched by stating that Christians are called to keep what God has created, including themselves, and implied: not create cloned human embryos.

15.5   The Theological-Ethical Argumentation

It may be worthwhile to reflect on what types of arguments churches and theologians have used in their ethical evaluation pro and contra the use of human embryonic stem cells in medical research and therapy.

An interesting classification of used and eventually useful arguments is delivered in a German publication by Damschen and Schönecker (2003). They distinguish between four main arguments which can be identified in the general, ethical debates concerning our topic. In their publication each of the arguments is given various formulations by the contributors (to the following version, p. V and 1 ff.). The first argument is called the *species* argument. In a simple version it may run like this: In the capacity of being members of the species *homo sapiens* the embryos are human beings and have dignity ('Würde') and shall be protected. The second argument is described as the *continuity* argument, which is given the following content: Embryos develop continuously without morally relevant stages into adult human beings who have dignity. The third is marked as the *identity* argument, stating: Embryos are in ethically relevant respects identical with adult human beings who have dignity. The fourth argument is called the *potentiality* argument defined in this way: Embryos have potential to grow into human beings and these potential humans are worth unlimited protection.

Here we have to keep in mind, however, that normative arguments in Christian ethics are not always directly compatible with corresponding normative elements in ethics belonging to other religions or various philosophical traditions. But if we use the four argumentative categories as analysing tools, we may still ask: Are there some similarities between these general arguments in the embryo-discussion and the ways of argumentation from the theological point of view reviewed above?

One obvious result is that the most common argument within Christianity contra the use of human embryonic stem cells in research, is the species argument. We have found it in the Orthodox Church and among Orthodox theologians, in the Roman Catholic Church and in several Protestant churches. But compared to a mainly biological concept of species it is in Christian ethics an anthropological concept 'human being' that is used. It has as its content the unique creature, brought into existence by an act of God, given life as a gift from him and having the right to live and not to be harmed, but the concept also includes knowledge from embryology. The meeting point between biology and theology in this context is the generally maintained view that being a 'human' in this sense is a reality from the time of fertilization. In common with the general embryo debate the argumentation from the side of churches and theology includes the aspect of certain rights belonging to the human life at this stage. But keeping in mind the theocentric and creational perspectives of this argument within Christianity, I prefer not to call it the species argument, but instead the argument of *specificity*.[8] However, the more precise, theological substance of the argument, below the references to creation and image of God, is rare deeply explained.

The argument of continuity can also be identified in the material analysed above, although it is not central or fundamental on the same level as the first one. But it is one of the reasons contra the use of embryonic stem cells in research delivered from the Orthodox as well as the Roman Catholic tradition. In all cases one refers to the continuous development of an embryo from the point of fertilization. The intention of using this argument is the rejection of any possibility of differentiation between stages with unequal moral status.

The identity argument I have not been able to find in the texts from churches.

But the potentiality argument is used in the statement from the Orthodox Church and the Presbyterian Church (USA) and among Roman Catholic theologians. The reference to potentiality gets its validity as defence for the status of an embryo as a human being deserving protection.

Leaving the four arguments from the German classification behind us, one may wonder what sorts of positive ethical reasons are additionally presented within Christianity. Some supplementary aspects must be mentioned. The heavy philosophical discussion about whether an embryo is a person or not, is not referred to by the churches. But the Roman Catholic Instruction *Donum Vitae* seems to apply the concept 'person' to embryos, as shown in several of the expressions quoted above. And the Orthodox Church in America ascribes an embryo 'personal existence' understood as 'ensouled existence'. In both cases being 'person' is inseparably linked with the specific status as human being from the moment of fertilization. And on the same line as other human 'persons' embryos have to be physically protected.

On both Orthodox, Roman Catholic and German Protestant sides we have discovered the argumentation that a good purpose (health for the sick) can not be ethically justified by means that are contrary to ethical standards (destructing the life of human embryos). This way of arguing regarding an ethical conflict between aims and means shows that the protection of the human life of embryos is given priority over against possible but uncertain therapeutic fruits. To put it in another way: Some defined deontological limits are evaluated as unalterable compared to the relevant teleological ends that might follow.

Of interest are also the efforts to rely upon the relevance and validity of the Hippocratic Oath (by the Orthodox Church in America) and a medical code and human rights texts (by the Southern Baptist Convention). In no cases is it reflected upon as to whether unborn humans are within the horizon of the documents referred to.

---

[8] On this point I am picking up a terminology from Gene Outka, see his contribution "The Ethics of Human Stem Cell Research", in ibid.: 29–64.

A part of one of the Orthodox statements argues against embryonic stem cell research by focusing on this particular kind of research. The Holy Synod of Bishops in the USA has several critical remarks to such science: the role of market forces, unrealistic promises with regard to therapeutic potential, a slippery slope tendency concerning respect for human life, absence of informed consent etc. These aspects of the research contribute to the ethical disqualification of it.

Arguments by churches pro research on human embryonic stem cells are to be found on different lines. One is the graduation of human development and its consequence in denying the individuality and personhood of human life at an early stage and relativizing its potential to grow to a human being. We have found elements of this view by Margaret A. Farley. This position is worked out more substantially by the nine Protestant theologians from Middle Europe.

A related way of thinking is the definition of an embryo as a 'growing human being' without exclusively fixing the starting point of this development to fertilization. This opens a possibility for differentiating between embryos who are given conditions for continuous growth until birth and embryos who are given limited time and are not supposed to be born. Embryos of the latter description are available for research. We have identified this position within a branch of the Evangelical Church in Germany.

Another argumentative line is a reference to the medical aims in the form of new possibilities of treating serious diseases. This teleological argument by focusing on lifesaving, health and freedom from suffering exists in the texts from the Evangelical Church in Germany, the Presbyterian Church (USA) and the United Church of Christ. But there is a characteristic difference in the material: In the German case the good purpose is confronted with and given priority behind the dignity and protection of human embryos. In the Presbyterian text the ethical dilemma between medical utility and respect for human life is solved by giving priority to potential, health care over against respect towards embryos on the condition that they are surplus. By the United Church of Christ no ethical dilemma is identified and the health argument is strengthened by a theological perspective: the healing service of Jesus as an example for the church.

What are then the positions of churches and theologians regarding different alternatives for eventually obtaining embryonic stem cells for research and therapy purposes? (a) A complete rejection of all kind of such research, independent of possible cell sources, is delivered by the Orthodox church, the Roman Catholic Church, the Council of the Evangelical Church in Germany, Southern Baptist Convention and United Methodist Church (focusing on cloning). This includes refusal of use of stem cell lines, surplus embryos, embryos produced for science and therapeutic and reproductive cloning. (b) Use of embryonic stem cell lines is accepted by the group of nine Protestant theologians. (c) Using surplus embryos after in vitro-fertilization is supported by Church of Sweden, the Presbyterian Church (USA) and the nine European theologians. (d) Production of human embryos for research purposes is rejected also by the nine theologians. (e) Therapeutic cloning is not rejected by the Church of Sweden and some members within a Commission of the Evangelical Church in Germany.

15.6   Conclusion

Summing up, the main result is that there is a broad agreement among the Orthodox church, the Roman Catholic Church and several Protestant churches that the use of embryonic human stem cells for research and therapy is totally unacceptable – seen from the point of Christian bioethics. Some Protestant churches like the Church of Sweden, the Presbyterian Church (USA) and the United Church of Christ to varying degrees evaluate such research and therapy positively.

What then seems to be the real burning questions in the theological discussion within Christianity? I shall focus on some characteristic problems: (a) *Human dignity* – is it original, is it gradual or is it final? And what is the sustainable, theological justification of it? (b) What is *personhood* and is it ethically relevant for the question of the protection of embryonic, human life and if so, in what way? (c) How does theology interpret and include *embryological knowledge* about unborn, human life in its anthropology? (d) What is the moral status of *medical research* according to a Christian understanding of creation, culture and human life conditions? (e) What weight do *ethical values* such as health, curing of diseases and reduction of suffering have when they are *uncertain*? (f) How does Christian bioethics deal with the *ethical dilemma* between medical purposes and the duty to establish limits for such research and therapy?

These are controversial questions which will continue to challenge churches and theologians also in the future.

References

Anselm, R. et al. (2002). "Pluralismus als Markzeichen. Eine Stellungnahme evangelischer Ethiker zur Debatte um die Embryonenforschung". *Frankfurter Allgemeine Zeitung*, Mittwoch, 23 January, Nr. 19: 8.

Breck, J. (1998). *The Sacred Gift of Life. Orthodox Christianity and Bioethics*. Crestwood, NY, St. Vladimir's Seminary Press.

Carlberg, A. (2004). "Kristna perspektiv på genetiken och dess tillämpningar". In: Andrén, C.-G. and Görman, U. (eds.). *Etik och genteknik. Filosofiska och religiösa perspektiv på genterapi, stamcellsforskning och kloning*. Lund, Sweden, Nordic Academic Press: 113–159.

Congregation for the Doctrine of the Faith (1987). *Instruction on Respect for Human Life in Its Origin and on the Dignity of Procreation (Donum Vitae)*. Rome, February 22. http://www.vatican.va/roman_curia/congregations/cfaith/documents/rc_con_cfaith_doc_19870222_respect-forhuman-life_en.html

Damschen, G. and Schönecker, D. (eds.) (2003). *Der moralische Status menschlicher Embryonen. Pro und contra Spezies-, Kontinuums-, Identitäts- und Potentialitätsargument*. Berlin/New York, Walter de Gruyter.

Demopulos, D. (2001). "A Parallel to the Care Given the Soul: An Orthodox View of Cloning and Related Technologies". In: Cole-Turner, R. (ed.). *Beyond Cloning. Religion and the Remaking of Humanity*. Harrisburg, PA, Trinity Press International: 124–136.

epd-Dokumentation Nr. 26/01 (2001). Evangelischer Pressedienst. Frankfurt am Main, 18 June.

Farley, M. A. (2001). "Roman Catholic Views on Research Involving Human Embryonic Stem Cells". In: Holland, S. et al. (eds.). *The Human Embryonic Stem Cell Debate. Science, Ethics, and Public Policy*. Cambridge, MA, MIT Press: 113–118.

Im Geist der Liebe mit dem Leben umgehen. Argumentationshilfe für aktuelle medizin- und bioe-thische Fragen. Ein Beitrag der Kammer für Öffentliche Verantwortung der Evangelischen Kirche in Deutschland. EKD-TEXTE Nr. 71. Kirchenamt der Evangelischen Kirche in Deutschland (EKD). Hannover 2002.

Kock, M. and Lehmann, K. (2002). *Zur Entscheidung des Deutschen Bundestages über den Import menschlicher embryonaler Stammzellen.* http://www.ekd.de/presse/397_pm9_2002_kock_lehmann_embryonenimport.html

Pontifical Academy for Life (2000). *Declaration on the Production and the Scientific and Therapeutic Use of Human Embryonic Stem Cells.* Vatican City, 25 August. http://www.vatican.va/roman_curia/pontifical_academies/acdlife/documents/rc_pa_acdlife_doc_20000824_cellule-staminali_en.html

Statement on the Ethical and Moral Implications of Stem Cell and Fetal Tissue Research. Actions of the 213th General Assembly (2001) from the Office of the General Assembly (the Presbyterian Church in USA). http://www.pcusa.org/oga/actions-of-213.htm

The Holy Synod of Bishops of the Orthodox Church in America (2001). *Embryonic Stem Cell Research in the Perspective of Orthodox Christianity.* New York, 17 October. http://www.oca.org/docs.asp?ID=50&SID=12

The Holy Synod of Bishops of the Orthodox Church in America (2002). *The Cloning of Human Embryos.* New York, January. http://www.oca.org/Docs.asp?ID=51&SID=12

Waters, B. and Cole-Turner, R. (eds.) (2003). *God and the Embryo. Religious Voices on Stem Cells and Cloning.* Washington, DC, Georgetown University Press.

Chapter 16
Theological Arguments in the Human Stem Cell Debate: A Critical Evaluation

Gunnar Heiene

Abstract The article focuses on the contribution of theology to the present debate on human stem cells. First, it discusses the role of theology and Christian ethics in a secular society with special regard to debates on bioethical issues in general and the human embryonic stem cell debate in particular. Secondly, it gives an overview of different argumentative strategies in theological contributions to bioethical issues. This overview is a background for an analysis of different theological arguments that have been used in the debate on stem cells, especially arguments in favour of or against the use of human embryonic stem cells. Four prominent participants in the American debate are presented, two Roman Catholics, Richard Doerflinger and Lisa Sowle Cahill, and two Protestants, Gilbert Meilaender and Ted Peters. The conclusion of the analysis is that specific theological viewpoints should be included in the debate on human embryonic stem cells in a way that could be understood by all participants in the debate in our pluralist society.

Keywords Theology and bioethics, stem cell ethics, Roman Catholic bioethics, Protestant bioethics, Christian viewpoints on stem cells

16.1 Introduction

The aim of this article is to present and analyze the contribution of theology to the present debate on human stem cells. First, I will discuss the role of theology and Christian ethics in a secular society with special regard to debates on bioethical issues in general and the human embryonic stem cell debate in particular. Then I will give an overview of different argumentative strategies in theological contributions to

MF Norwegian School of Theology, P.O. Box 5144, Majorstuen, 0302 Oslo, Norway.
e-mail: Gunnar.Heiene@mf.no

bioethical issues. This overview is a background for my main point, where I will present and analyze different theological arguments that have been used in the debate on stem cells, especially arguments in favour of or against the use of human embryonic stem cells. Finally, I will draw some conclusions from the critical analysis of the theological arguments.

16.2 The Role of Theological Arguments in Debates on Bioethical Issues

In the modern secular society, the role of theological arguments in debates on common ethical challenges is disputed. On the one hand, there is a fear that a 'stem cell theology' should lead to very restrictive policy, on the other hand there are signs of a more open attitude to contributions from religious and theological perspectives.

The fear of 'stem cell theology' is related to the present debate in the United States, where President George W. Bush and other conservative republicans have used theological arguments to defend a restrictive view. In May 2005, House Republican leader Tom DeLay, in opposing a bill that would allow discarded early embryos to be used as sources of stem cells, presented an argument with an unmistakable religious undertone: 'An embryo is a person, a distinct internally directed self-integrating human organism. We were all at one time embryos ourselves. So was Abraham. So was Muhammad. So was Jesus of Nazareth.' A Jewish professor at Harvard Medical School, Jerome Groopman (2005), criticized this argument, claiming that it is 'foolish, and wrong, to use the founders of Judaism, Islam and Christianity as foils to support the current administration's view on pressing moral questions in medicine'. Groopman was not the only one to raise a critical voice. In an editorial article, *New York Times* (2005) commented on the president's stem cell theology: 'His actions are based on strong religious beliefs on the part of some conservative Christians, and presumably the president himself. Such convictions deserve respect, but it is wrong to impose them on this pluralistic nation.' President Bush threatened to veto the bill that was approved by the House of Representatives, claiming that a bill that would encourage the destruction of embryos from which the stem cells are extracted, would 'take us across a critical ethical line'. *New York Times* claimed that this was a 'powerful proof of the dangers of letting one group's religious views dictate national policy'.

The same critical comments have followed upon President Bush's second veto in less than a year against the bill expanding embryonic stem cell research 20 June 2007. According to *New York Times* (2007), the veto aims at preventing research that would involve 'the destruction of microscopic entities – smaller than the period at the end of this sentence – that the president deems a nascent form of life'. This is regrettable, according to the editorial article, since almost all scientists in the field consider embryonic stem cell research the most promising: 'It is foolish to crimp that research by withholding federal funds to placate a minority of religious and social conservatives, including Bush, who deem the work unethical.'

The critical attitude towards the President's 'stem cell theology' may be seen as an expression of the development characterized by Daniel Callahan (1990) as 'the secularization of bioethics'. According to Callahan, the most striking change in bioethics the last decades has been the move from a bioethical position dominated by religious and medical traditions to a position dominated by philosophical and legal concepts. In public discourse, secular themes like universal rights, individual self-direction and procedural justice have dominated, at the expense of concepts of the common good or a transcendent individual good. Alluding to his personal life story, Callahan asks: 'Just as I had found I did not need religion for my personal life, why should biomedicine need it for its collective moral life?'

On the other hand, Callahan himself is not satisfied with this process of secularization. Biomedical ethics, he claims, cannot be said to be 'demonstrably more robust and satisfying as a result of its abandonment of religion'. But the loss of religion need not be the end of the story. 'Can those of us who share my lack of belief still make use of at least some of the insights and perspectives of religion, even as we reject its roots?', Callahan asks in his contribution 18 years ago. Still he is ambiguous in his evaluation of religious contributions. In a recent article on stem cell research he is still critical to the religious arguments in the debate, claiming that they don't get very far with those outside of some religious traditions. 'Proof-text fundamentalism on the Protestant side, or papal authority on the Catholic side, has little weight', he says (Callahan 2005). Personally, he still prefers a nonreligious way of making an argument against embryonic stem cell research.

The fear of a fundamentalist 'stem cell theology' should not prevent us from seeing the positive contributions to the present day debates in bioethics from religious and theological traditions. While criticising the conservative republican leaders' use of religion in opposing the bill, Jerome Groopman (2005) still admitted that religion and theology could have a positive role, which is often too quickly dismissed by secular scientists. 'The Bible and its commentaries are a wealth not only of ethical imperatives but also of insights into character and behaviour. It is foolish or naïve to ignore this fact', he claimed.

Callahan and Groopman are not the only ones to call for a religious and theological contribution to the debate on stem cells. In an article in *Nature*, Tony Reichhardt (2004) explores the relationship between science and religion on the background of the debate on embryonic stem cell research. At least one thing has changed in the debate between science and religion since Galileo's days, he remarks: 'Now, science is the orthodox worldview, in the industrialized world at least, and religion stands outside, raising objections.' Reichhardt quotes sociologist of religion, John Evans, who has noticed that at bioethics conferences, biologists rarely show any knowledge of theology, while 'religious people are expected to have spent huge amounts of time learning all the science'.

In 1990, Callahan's analysis of the secularization of bioethics was in accordance with the main paradigm of secularization. Recently, we have noticed a shift of paradigm. In his recent article 'Religion in the public sphere', Jürgen Habermas (2006:1) starts with the following comment: 'Religious traditions and communities of faith have gained a new, hitherto unexpected political importance since the epochmaking change of 1989–90.' Noting the political revitalization of religion in the heart of Western society, Habermas claims that the liberal conception of democratic citizenship based on the assumption of a common human reason is challenged and should be loosened. Although Habermas does not agree with Nicolas Woltersdorff, when he claims that religious arguments should be used by political legislatures, he supports Woltersdorff's general view that religious people ought to base their decisions concerning fundamental issues of justice on their religious conviction. Habermas admits that religious citizens do not need to 'split their identity into a public and a private part the moment they participate in public discourses. They should therefore be allowed to express and justify their convictions in a religious language if they cannot find secular "translations" for them' (Habermas 2006:10). Even secular citizens or citizens of a different faith could learn something valuable from religious citizens who publicly express and justify their convictions by resorting to religious language. The liberal state should not cut itself off from key resources that could come from religious traditions.

But there is also a need of translating from the religious 'first order' language to the common 'second order' language:

> Religious traditions have a special power to articulate moral intuitions, especially with regard to vulnerable forms of communal life. In the event of the corresponding political debates, this potential makes religious speech a serious candidate to transporting possible truth contents, which can then be translated from the vocabulary of a particular religious community into a generally accessible language. (Habermas 2006:10)

As far as I can see, this means that Habermas welcomes religious and theological arguments in the political public sphere, as long as these arguments can be formulated and justified in a language that is equally accessible to all citizens. This involves a 'modernization of religious consciousness' as a response to the challenges of pluralism, modern science and the spread of positive law and a profane morality. The work of hermeneutic self-reflection within religious traditions has essentially been performed by theology. The question is unanswered, however, whether the proposed concept of citizenship 'in fact imposes an asymmetrical burden on religious traditions and religious communities after all' (Habermas 2006:14). Therefore, also secular citizens should be challenged in the same way as religious. In our society, complementary learning processes are required from both religious and secular citizens. Not only do religious people need to understand the secular mind; secular citizens have to learn what it means to live in a post-secular society. For his own part, Habermas outlines 'a post-metaphysical thought' as the secular counterpart to a religious consciousness that has become self-reflective, rejecting a narrow scientistic conception of reason. This means that it is possible to discover the internal relationship not only to the classical philosophical traditions, but also to the world religions, especially to the religious traditions that have influenced our culture.

Habermas' 'post-metaphysical' position is a challenge to theology. On the one hand, it invites theology to express arguments in public debates based on deeply rooted religious convictions, claiming that such arguments could be of value to all participants in the debate. On the other hand, Habermas expects theology to take part in a process of hermeneutic self-reflection and of translation, which arises from a reconstruction of sacred truths in the light of modern living conditions. This is a challenge also for Christian ethics in its attempts to develop sustainable arguments in the recent debates on bioethics, as in the controversy on the use of human embryonic stem cells. Theological arguments should be considered as interesting and potentially valuable not only for Christians and other religious persons, but for society as a whole.

16.3   Different Argumentative Strategies in Theological Contributions to Bioethics

In his book on theology and bioethics in the Nordic countries, Kees van Kooten Niekerk (1994) distinguishes between two main argumentative strategies in theological contributions within Protestantism. On the one hand, there are specific Christian arguments, based on theological concepts: the world as God's creation, man as created in the image of God and loved by God, neighbourly love, the fifth commandment, etc. On the other hand, many theologians prefer general arguments which could be accepted by all citizens, religious believers or not. Such arguments are: the intrinsic value of the living nature, the special value of human life and of man as person, the quality of life, beneficence and protection of the weak, etc. Niekerk shows that many theological participants in the public debate combine the two argumentative strategies, using both general and specific Christian arguments in bioethical issues. As an example, the Danish theologian Viggo Mortensen combines general arguments, based on K.E. Løgstrup's phenomenological analysis of human life, with a specific theological contribution, based on the resources which are unique for theology: (1) the Christian world view and Christian anthropology, (2) the Biblical narratives, giving witness to a specific way of treating human problems, (3) the commandment to love one's neighbour and the ethical pattern of argumentation within the theological tradition, and (4) Jesus as example.

Looking at the recent contributions from theology and Christian ethics in the debate on stem cell research we see the same pattern as reported by Niekerk. Some theologians prefer to argue strictly on general, secular premises, while others prefer arguments drawn from the specific Christian tradition, referring to the Bible and the Christian tradition. A combination of these two argumentative strategies is not unusual, starting either with the specific Christian contribution to the issue, giving more general arguments to support a position already taken, or starting with general arguments, then turning to arguments from the religious tradition as a specific contribution to the issue of stem cell research.

In an article with the title 'Ethical Issues: A Secular Perspective', Professor Ernlé W.D. Young (2001:163) starts asking why someone trained in theological ethics should find it necessary to 'comment from a secular perspective on the moral standing of human embryonic stem cells and germ cells'. He answers by claiming that there is an important difference between 'morality' as tied to religious traditions, and 'ethics', understood as a 'common or public language in justifying assertions about prescribed or proscribed attitudes and actions'. Justifications deriving from specific religious traditions are regarded as a 'private' language, to be understood only by adherents of the same tradition, while 'the more neutral language of reason' is the only way to make moral arguments in a public forum, according to Young. He is quite pessimistic in his analysis of the differences between ethics and morality. While ethics depends heavily on reason, and is more open to uncertainties and ambiguities, morality based on deeply held beliefs, values and convictions tends to leave little or no room for compromise, thinking more in categories of right and wrong, black and white, Young claims.

Young's position is a rejection of the value of theological arguments based on a specific religious tradition in the stem cell debate. For Young, there is a fundamental difference between arguments based on reason and on faith. This is not the case for all theologians who prefer to use a general language. Especially from the Roman Catholic side, using general arguments appealing to human reason has been a common strategy. This strategy is of course based on the natural law tradition. As John Langan (2003) comments in an article on stem cell research, catholic doctrine in this area 'rests on rationally accessible propositions about the substantial continuity of the person from conception on to maturity and about the respect that should be shown for innocent human life in general'. According to Langan, this view does not rely on religiously idiosyncratic premises and it does not require the gift of faith for its affirmation. For this reason, Langan, and many other Roman Catholic contributors to the stem cell debate take Thomas Aquinas' natural law concept as a starting point. According to Langan, Aquinas thinks that natural law precepts are found on three levels of specificity and knowability, (a) highly generic precepts such as "Good is to be done, evil is to be avoided' and the love command, (b) specific precepts corresponding to the natural inclinations of all humans, forbidding certain kinds of actions, e.g., the prohibition of killing in the Ten Commandments, and (c) further conclusions and specifications of these precepts. To reach such conclusions, for example a negative judgment on stem cell research, one needs additional premises about the development of human life.

In Protestant contributions, there is a tendency to use arguments specifically linked to the Bible and the Christian tradition. In an article titled 'Some Protestant Reflections', Gilbert Meilaender (2001), one of the main representatives of a restrictive attitude in the stem cell debate, develops his argument from reflections in the writings of Karl Barth, John H. Yoder and Stanley Hauerwas. This means that arguments about the personhood of Jesus Christ and the nature of the Christian community play an important role. A more liberal protestant, Ted Peters, also gives specific Christian arguments a prominent place, utilizing the eschatological character of Christian faith as a key. The distinctive feature of Christian anthropology is its orientation toward the future. This means that proleptically, humans participate in the grace of God's eschatological destiny. 'Ethically speaking, grace appears to reorder reason', Peters claims (Peters and Bennett 2003). Although this article is inspired by K.E. Løgstrup's ethics, it says that Christian anthropology should deepen the understanding of human ethics from a specific Christian understanding of what it means to be a human being.

But Peters and many other Protestant contributors do not only use specific Christian arguments. Often such arguments are combined with general arguments on a common morality. In a statement from the Southern Baptist Convention (2003) on stem cell research, the starting point is the Biblical teaching of human beings as made in the image and likeness of God. After a sharp criticism of a 'crass utilitarian ethic which would sacrifice the lives of the few for the benefits of the many', the statement uses general arguments about consequences and duties. Another example of this combination of specific Christian and more general arguments is a statement from the Orthodox Church in America (2003), starting with the conviction that human life begins at conception, a conviction 'grounded in the Biblical witness' (references to Ps 139:134–16; Isaiah 49:1 ff.; Luke 1,41,44) as well 'in the scientifically established fact that from conception there exists genetic uniqueness and cellular differentiation that, if the conceptus is allowed to develop normally, will produce a live human being'. The arguments are of a general kind, such as negative consequences, the slippery slope argument, the subject's informed consent, etc.

Although specific Christian perspectives play an important role, either as a starting point or as a road to a deeper understanding of the common human situation, the general arguments based on a common morality take a prominent place in most theological contributions to the stem cell debate. Obviously, this has to do with the complexity of the issue and the need to be understood in the public arena. On the other hand, most writers think it is necessary to present the specific religious contributions from a Christian theological perspective. This speaks for a 'combined' model as the best argumentative strategy from Christian theology in discussing stem cell research and the use of human embryonic stem cells in treatment.

16.4   Theological Arguments in the Stem Cell Debate: Differences and Uniting Perspectives

We have seen that theologians prefer different argumentative strategies, although a combination of general arguments and specific Christian arguments is the most typical strategy. Our next question is about the content of the arguments given from a theological perspective. There is certainly not full agreement among theologians in the issue of stem cells, and to show some of the important differences, I will present four prominent participants in the American debate, two Roman Catholics, Richard Doerflinger and Lisa Sowle Cahill, and two Protestants, Gilbert Meilaender and Ted Peters.

16.4.1   A Conservative Roman Catholic Voice: Richard M. Doerflinger

Doerflinger's starting point in his presentation of a 'catholic viewpoint' is the official view on the moral status of the human embryo in Roman Catholic doctrine (Doerflinger 1999).

Human life is seen as a continuum, and human individuals of every age and condition are seen as meriting the same respect for their fundamental right to life. Building on new biological knowledge about the early embryo, he rejects the view presented also by some Catholic thinkers that there is a qualitative difference between the 'pre-embryo' and the embryo after 14 days. The role of the term 'pre-embryo' is criticized, and Doerflinger welcomes the recent dismissal of this term.

Another argument that has been used in favour of embryonic stem cell research, is the argument about 'twinning'. While biologists earlier used to see the early embryo until the appearance of the 'primitive streak' as a largely formless mass of interchangeable and undifferentiated cells – capable of splitting into two or more embryos (hence without inherent individuality) – it now seems, Doerflinger says, that an embryo's potential for spontaneous 'twinning' is established very early, so that the vast majority of embryos, from the outset, do not have the property of producing twins spontaneously. Besides, new discoveries suggesting the possibility of human cloning also from adult cells may adjust our perceptions of the embryo, Doerflinger says, claiming that 'each new finding has underscored the Catholic view of human life as a continuum from the one-cell stage onward' (Doerflinger 1999:139).

Still, many people argue that an embryo can't be given the moral status of a person, since it lacks various mental or physical abilities. According to Doerflinger (1999:139), all versions of this argument face a dilemma:

> Either the ability in question is an 'either/or' proposition, so that some arbitrary level of functioning must be stipulated to divide 'persons' from 'nonpersons', or it admits of degrees, in which case 'personhood' must admit of degrees as well.

Neither should personhood be viewed as a 'social contract', as proposed by Ronald M. Green. This view implies that there is nothing inherent in any human being that obliges us to respect that being as a person. The 'social contract' view of personhood, understood as an attempt at a coherent moral defence for destructive human embryo experiments, could have serious consequences, since it offers a rationale for conducting harmful experiments on human subjects after birth as well.

Doerflinger also comments on the issue of complicity, which has played an important part in Roman Catholic contributions both on the use of foetal tissue from induced abortions and (after President Bush's famous speech August 2001) on the use of stem cell lines from embryos which had already been established. According to Doerflinger, use of material from aborted foetuses is not rejected in principle, but must meet 'moral requirements', such as 'no complicity in deliberate abortion'. But the same is not true of the new proposals for destructive harvesting of stem cells from living embryos, Doerflinger claims, since those who harvest and use the cells are necessarily complicit in the destruction of the embryo. Commenting on the argument that 'spare' embryos that are not longer needed for reproductive purposes will be discarded anyway, Doerflinger claims that harvesting cells from such embryos is very different from harvesting tissues when that individual is already dead.

Another challenge is the question whether it is morally wrong not to use a medically beneficial resource, which could have been developed from harvesting cells from human embryos. Doerflinger's first argument against this view is that harmful experiments never must be performed on the embryo unless they are the only feasible means for obtaining vitally important medical benefits. As research on adult stem cells has shown to be promising, and in some respects even more clinically useful than embryonic cells, it is impossible to argue with harvesting stem cells from embryos as a last resort. Besides, treatment with stem cells from embryos could create problems of conscience for many future patients, Doerflinger claims. According to Catholic social teaching, society's resources should be used to promote the common good, with a preference for the weakest and most vulnerable. This means that all human embryos should be protected from destructive experimentation, regardless of the source of funding. Doerflinger admits that in a pluralist society there are limits for what could be labelled as unlawful, but still it is important that such activities are not supported by governmental resources.

On the basis of the Roman Catholic natural law tradition, Doerflinger's conservative view on the status of the human embryo, the issues of complicity and of federal funding of stem cell research are developed with arguments which could be held also by people not sharing the Christian faith. Still, Doerflinger underlines the specific contributions to the debate from a Catholic perspective, including the realism of the Catholic world view, knowing that we do not control the world. Besides, he points to the eternal consequences of wrong moral decisions, the 'common good' tradition and the defence of rights of conscience as specific Catholic contributions to the stem cell debate.

Doerflinger's arguments are presented as a coherent set of propositions, formulated in a general language that can be understood by people from different faiths and world views. Still, there is a problem in his argument, because of its strong dependence upon recent scientific developments. There is no guarantee that biologists will produce insights that could support Doerflinger's arguments about embryonic development and the problem of 'twinning', although his case in rejecting the term 'pre-embryo' obviously finds support in recent scientific developments.

16.4.2   A Moderate Roman Catholic Voice: Lisa Sowle Cahill

Lisa Sowle Cahill (2003) represents a moderate position within the Catholic tradition, accepting in principle embryonic stem cell research, but with many reservations. This research should be done very cautiously and 'in a spirit of great moral reservation', she argues. She concludes that the research use of embryos that are to be discarded in any event may be justifiable, although this is not unproblematic.

Her argument is based on five moral values, woven together as five cords. The first is *the value of nascent life*. This has to do with the moral status of the embryo. She rejects a view on the embryo as merely a 'clump of cells'. On the other hand the official documents from the Roman Catholic Church asserts that the earliest embryo is to be considered as a person. Her own preferences lie somewhere in the middle, building on the view expressed by Richard A. McCormick that the embryo could be referred to as 'nascent human life', as a developing life whose value is not merely potential but still not fully realized human life until later. The 'developmental' view has been important among more moderate and liberal catholic ethicists, stating that the early human embryo is worthy of respect, although personal moral status should not be given until a certain degree of development has taken place. Also in official documents like the 1987 Vatican *Instruction on Respect for Human Life in Its Origin and on the Dignity of Procreation*, she finds a more nuanced position on the personal status of the embryo. On the one hand, the document claims that 'the human being is to be treated as a person from the moment of conception', on the other hand the document is vague on the issue of the presence of the human soul in the embryo. According to Cahill, the full personal status of the early embryo is so doubtful that the case should be solved by traditional Catholic ways of solving such problems. First, probabilism will be helpful to move from uncertainty to a choice about the right action. This means that common sense is given an important position in the moral argument. Probabilism allows for a limited use of human embryos in stem cell research without necessarily rejecting the general principle that it is wrong directly to kill an innocent human person, since there is a doubt concerning the status of the early embryo. According to probabilism, it is permissible to adopt an opinion if it is probable, even if the opposite is more probable, Cahill argues.

The second moral value proposed by Cahill is *the value of moral virtue*. In Catholic moral theology the issue of complicity and cooperation has been discussed thoroughly. Doerflinger also reflects on this issue in his argument against embryonic stem cell research, but according to Cahill the complicity question is not in itself definitive in this case.

The value of medical benefits is Cahill's third value. The basic ethical principle of beneficience implies that we should serve human health through scientific and technological developments. Within Catholic moral theology, the principle of double effect has been used as a tool to decide whether it is permissible or not to cause some evil in the pursuit of the good. But this principle should not be used as a simple utilitarian principle of 'the greatest good for the greatest number'. The decisive question is whether the human embryos which are destroyed, are persons or not. If not, destroying them is not necessarily intrinsically evil, Cahill argues.

The value of distributive justice, implied in the principle of the common good, also has to be recognized in the stem cell debate. This could be a critical argument against the present stem cell research. In a global perspective the regeneration of tissue by stem cell techniques is 'as exotic as it is expensive', Cahill comments.

The last value proposed by Cahill is *the value of a social ethos of generosity and solidarity*. Cahill draws attention to John Paul II and others adopting the principle of a 'preferential option for the poor'. Solidarity is the most important social virtue, and in *Evangelium Vitae* John Paul II rejects a complete individualistic concept of freedom. This value supports a critical attitude toward developing profitable medical miracles for the elite. Therefore, the market in biotechnology and medicine should be limited by a sense of the common good, mutual rights and duties, and the participation of all in goods to be shared.

Taken together as five cords woven together, Cahill's argument is more suggestive than conclusive. She combines a critical attitude towards embryonic stem cell research with openness to the use of leftover IVF embryos. She admits that this is a complicated and ambiguous question; still she claims that the Catholic tradition provides tools to guide moral decision making in difficult questions like stem cell research. In a Catholic contribution, concerns about the embryo should be framed within a larger call for 'respect for human dignity', she claims, quoting John Paul II. Cahill's contribution is a well balanced argument, taking all the different aspects of stem cell research into consideration. The difficult point is her argument based on probabilism, claiming that there is reasonable doubt about the moral status of the early embryo. This is the reason why she admits a limited possibility for using embryos in stem cell research.

16.4.3   A Conservative Protestant Voice: Gilbert Meilaender

In his 'protestant reflections' on stem cell research, Gilbert Meilaender (2001) admits the 'complexities' of the issue. His starting point is a sentence from Karl Barth on the importance of the community's protection of its weak and even its very weakest members. On this background, Meilaender discusses the moral status of the embryo. As opposed to the present tendency to distinguish between persons and human beings, Meilaender introduces the traditional Christian concept of the person in the framework of Christology. In order not to be misunderstood so that Jesus of Nazareth could be seen as two individual persons, identified in terms of two sets of personal capacities, Christian theology went in another direction. A person is not someone with a certain set of capacities, but simply someone 'who has a history':

> The story, for each of us, begins before we are conscious of it, and, for many of us, may continue after we have lost consciousness of it, even when we lack or have lost certain capacities characteristic of the species. (Meilaender 2001:143)

In the light of the mystery of the human person and of our own individuality, there is an openness to honour the dignity of even the weakest of living human beings. This is Meilaender's basic argument in favour of the personal status of the human embryo.

Secondly, Meilaender warns against 'overselling' the promises and possibilities in stem cell research. We should learn to stop and to force ourselves to look for other possible ways to achieve desirable ends, for example, by using adult stem cells. It is important always to look for better solutions when the situation is ambiguous.

Thirdly, Meilaender quotes theological ethicist Stanley Hauerwas, who says that the church's primary mission 'is to be a community that keeps alive the language and narrative necessary to form lives in a truthful manner'. This means that it is important for the church to speak truly and straightforwardly, to avoid euphemism and equivocation in speaking about difficult questions like embryonic stem cell research. We should not deceive ourselves by supposing that we will use only 'excess' embryos from infertility treatments, having already created far more embryos than are actually needed for the treatment, and we should not use terms like the 'preembryo' or the 'preimplantation embryo', using terminology to diminish respect for embryos.

Meilaender's reflections, which are adapted from his testimony before National Bioethics Advisory Commission in May 1999, is an example of a conservative Protestant position, based primarily on specific Christian arguments, but at the same time with an openness to a general argumentative strategy. Still, we should ask whether the argument from Christology is sufficient to bear the weight of the argument for viewing the human embryo as a person.

16.4.4   A Liberal Protestant Voice: Ted Peters

Generally speaking, a liberal position on embryonic stem cell research is more common among Protestant theologians than among their Catholic colleagues. One of the main critics of traditional Catholic moral theology in this case (and also of conservative Protestant positions), is Ted Peters (2003). He claims that the traditional theological arguments about the early embryo as a person and about the ensoulment of the embryo are insufficient, since they imply nonmalificence as the prominent ethical principle. The result is that other relevant theological considerations are ignored, Peters claims. The debate has concentrated too much on the derivation question, i.e. the question of where we get stem cells from, with the protection of the early embryo as the overriding concern. The result of this narrow scope is that 'nonmalificence forcefully trumps beneficence' (Peters 2003:58). Not only do we see this in conservative churches as the Southern Baptist Convention; even in the more liberal United Methodist Church we observe virtually the same restrictive position; not to speak about the Vatican position, Peters complains. He criticizes Vatican creationism in the understanding of the human soul, especially as ensoulment is tied to genetic uniqueness, as in *Donum Vitae* and *Evangelium Vitae*. The tie between DNA at conception and the establishment of an individual human being is simply a false assumption, according to Peters: The cells of the early embryo do not become an identifiable individual human being until they adhere to the uterine wall at about 12 to 14 days. A first reading of Vatican statement would be that morally protectable personhood is dependent upon infusion of the spiritual soul into the physical body, but a closer look will show that *fertilization* is the decisive point in establishing personhood. A problem here is how the church finally could justify protecting the dignity and personhood of twins and clones, and besides, the Vatican position doesn't give the precise connection between ensoulment and personhood.

Peters' own alternative is an argument based on a theological anthropology stressing the relational character of human beings. Genetic uniqueness is not in itself a measure of personhood, dignity, or moral protectability, Peters claims. This view is individualistic in its essence and should be corrected by a more relational view: 'Biological uniqueness does not imply independence; we are who we are because of our relationships.' (Peters 2003:70). From a relational concept of personhood Peters understands dignity as 'the result of grace', of God's divine love. Unlike Meilaender, Peters uses the concept of person developed in the Christian tradition as an argument for a more liberal attitude to stem cell research. Personhood requires communion and cannot be understood merely in biological terms, Peters claims.

Closely linked to this argument is the eschatological argument, saying that personhood and dignity is proleptic, future oriented. Our 'dignity as human beings derives more from destiny than from origin, more from our future than from our past'(Peters 2003:71). The essential human nature is found not in Adam, but in Christ, Peters claims, with reference to Karl Barth.

Peters' argument is a theological argument, with a critical attitude especially to the Roman Catholic arguments, which he claims are based more on biology than on theology. Still, critical questions could be asked also to Peters' position. Is there a necessary link between fundamental concepts in theological anthropology like 'person in communion' and 'person as proleptic' and Peters' liberal view of embryonic stem cell research? Some of Peters' critical comments, especially on the Catholic moral tradition, are well worth listening to, but still it is questionable whether or not he has presented an argument that could solve the debate on stem cell research from a theological perspective. The question of the moral status of the embryo is fundamental to the theological debate, and as shown by Cahill, a focus on this question does not mean that other relevant perspectives are excluded from the debate.

16.5 Conclusion

We have seen that theological arguments on stem cell research differ considerably, both in argumentative strategy and in content. Still, important differences between a Catholic and a Protestant tradition can be traced, but there are also examples of a more common understanding of the issue. In my view, all the four contributions presented could shed some light on the stem cell debate. However, arguments which tend to isolate theology from biology, are problematic. Theological ethics should develop arguments based in the specific Christian contribution, and at the same time be willing to have an ongoing dialogue with biologists and other experts in the field. The strength of the theological contribution would then be its overall perspective on the issue of stem cell research, concentrating not only on the issue of the moral status of the embryo, but on all morally relevant perspectives on stem cell research.

In the further debate, theological arguments should reflect more on what it means for humans to be 'created co-creator', underlining the ambiguous nature of human creativity (cf. Hansen and Schotsmans 2001). On the one hand, we should apply our technological ingenuity to reduce human suffering, on the other hand, we should be well aware of the tendency in modern society to view human life as merely objects, means or functions. In the stem cell debate we are dealing with a conflict of important values. As stated by Lisa Sowle Cahill (2001:19), there can be reasonable arguments both for and against human embryonic stem cell research, but still theology needs to 'supply a corrective to the idea that new biomedical techniques are an unbounded and unassailable force for good'. The task of theology is to bring out the specific theological viewpoints that should be included in the debate on human embryonic stem cells, and at the same time present them in a way that could be understood by all participants in the debate in our pluralist society.

References

Cahill, Lisa Sowle (2001). "Stem Cells: A Bioethical Balancing Act", *America*, March 26:14–19.

Cahill, Lisa Sowle (2003). "Stem Cells and Social Ethics: Some Catholic Contributions". In: Nancy E. Snow (ed.). *Stem Cell Research: New Frontiers in Science and Ethics*, Notre Dame, IN: University of Notre Dame, pp. 121–142.

Callahan, Daniel (1990). "Religion and the Secularization of Bioethics", *The Hastings Center Report*, July/August:2–4.

Callahan, Daniel (2005). "Promises, Promises. Is embryonic stem-cell research sound public policy?", *Commonweal*, January 14:12–14.

Doerflinger, Richard M. (1999). "The Ethics of Funding Embryonic Stem Cell Research: A Catholic Viewpoint", *Kennedy Institute of Ethics Journal*, 9 (2):137–150.

Groopman, Jerome (2005). "Beware of Stem Cell Theology", *Washington Post*, May 29:B07.

Habermas, Jürgen (2006). "Religion in the public sphere", *European Journal of Philosophy*, 14 (1):1–25.

Hansen, Bart and Schotsmans, Paul (2001). "Cloning: The Human as Created Co-Creator?", *Ethical Perspectives*, 8 (2):75–87.

Langan, John S.J. (2003). "Stem Cell Research and Religious Freedom". In: Nancy E. Snow (ed.). *Stem Cell Research: New Frontiers in Science and Ethics*, Notre Dame, IN: University of Notre Dame, pp. 37–46.

Meilaender, Gilbert (2001). "Some Protestant Reflections". In: Suzanne Holland et al. (eds.). *The Human Embryonic Stem Cell Debate: Science, Ethics, and Public Policy*, Cambridge, MA: MIT, pp. 141–147.

New York Times (2005). "The President's Stem Cell Theology", *New York Times*, May 26.

New York Times (2007). "Mr. Bush's Stem Cell Diversion", *New York Times*, June 21.

Niekerk, Kees van Kooten (1994). *Teologi og bioetik. Den protestantisk-teologiske vurdering af bioteknologien i Norden 1972–1991*, Århus, Denmark: Aarhus Universitetsforlag.

Orthodox Church in America (2003). "Embryonic Stem Cell Research in the Perspective of Orthodox Christianity: A Statement of the Holy Synod of Bishops of the Orthodox Church in America". In: Brent Waters and Ronald Cole-Turner (eds.). *God and the Embryo: Religious Voices on Stem Cells and Cloning*, Washington, DC: Georgetown University Press, pp. 172–176.

Peters, Ted (2003). "Embryonic Persons in the Cloning and Stem Cell Debates", *Theology and Science*, 1 (1):51–77.

Peters, Ted and Bennett, Gaymond (2003). "A Plea for Beneficience: Reframing the Embryo Debate". In: Brent Waters and Ronald Cole-Turner (eds.). *God and the Embryo: Religious*

Voices on Stem Cells and Cloning, Washington, DC: Georgetown University Press, pp. 111–130.

Southern Baptist Convention (2003). "Resolution: On Human Embryonic and Stem Cell Research". In: Brent Waters and Ronald Cole-Turner (eds.). *God and the Embryo: Religious Voices on Stem Cells and Cloning*, Washington, DC: Georgetown University Press, pp. 178–179.

Reichhardt, Tony (2004). "Studies of faith", *Nature*, 432, December 9:666–669.

Young, Ernlé W.D. (2001). "Ethical Issues: A Secular Perspective". In: Suzanne Holland et al. (eds.). *The Human Embryonic Stem Cell Debate: Science, Ethics, and Public Policy*, Cambridge, MA: MIT, pp. 163–174.

Chapter 17
Human Embryos and Embryonic Stem Cells – Ethical Aspects

Monika Bobbert

Abstract Research using human embryos and embryonic stem cells is viewed as important for various reasons. Apart from questions concerning legal regulations, numerous ethical objections are raised pertaining to the use of surplus embryos from reproductive medicine as well as the creation of embryos and stem cells through cloning. The question of the moral status of human embryos is not the only relevant one. Also other aspects as a critical view on the model of contract law, including parent's property rights on embryos, women's well-being and certain changes of context – e.g. the transfer of embryos from reproduction to research – must be taken into consideration. Moreover, research with morally sensitive goods is often balanced with high research goals. Some criteria are given, which a severe evaluation of value and reachability of research must suffice.

Keywords Human embryonic stem cells, human embryos, research goals, stem cell research, therapeutic cloning

17.1 Autonomous Morality Within the Context of Christian Faith: Remarks About the Relationship of Philosophical Ethics and Theology

Within catholic theology in German speaking countries the so-called model of autonomous morality within the context of Christian faith is widely accepted (Auer 1971). This means, to regard morality as being based on human freedom and

Section Medical Ethics at the Institute for History of Medicine (Bereich Medizinethik am Institut für Geschichte der Medizin), Medical Faculty, University of Heidelberg (Medizinische Fakultät der Universität Heidelberg), Im Neuenheimer Feld 327, 1. OG, D – 69120, Heidelberg, Germany.
e-mail: Monika.Bobbert@histmed.uni-heidelberg.de

responsibility. Kant's self-legislation is interpreted as divine liberation. Thus, moral autonomy is a gift of God. As a theory of rational ethics the autonomous morality within the context of Christian faith follows the tradition of natural law: The morally right can be discovered and understood by rational ways. Justification of ethical assessments can and must be given by philosophical means.

The Christian context is relevant for becoming sensitive and for discovering ethical problems. It is also relevant for the motivation to act in a responsible and morally reflected way. Religious motives and experiences can inspire ethical reflection, they can influence the gravity of ethical arguments. Beside this there are convergences between theological and philosophical concepts, e.g. between the belief that men is a creation of God and thus every person being unique und accepted un-conditionally and the humanistic thought that every human being has value in itself and must be respected and supported, irrespectively of his or her value for others. Another convergence can be seen in regard to human mortality and contingency. Existential insights in the possibility of errors and deficiencies of every person can be ethically relevant in discussions, e.g. about research with human beings, reproductive medicine or about the extent of medical means at the end of life.

Theological ethics in the sense of autonomous morality tries to integrate fundamental Christian beliefs into ethical reflection. But religious beliefs and persuasions cannot substitute ethical argumentation. Moreover, the model of autonomous morality within Christian Context aims towards an ethics which can be communicated and generalized. This approach attempts to reformulate fundamental norms and concepts in philosophical terms and arguments. Although in principle no special way of philosophical reasoning is given an advantage certain ethical approaches seem hardly to consent with Christian faith, for example hedonism, pure utilitarianism or determinism. One way of a rational foundation for the idea of human dignity and uniqueness of every person is the ethical theory of the philosopher Alan Gewirth, who stands in the Kantian line (Gewirth 1978). He provides grounds for individual moral rights as well as criteria for weighing conflicts of rights.

Theological ethics as applied ethics is necessarily an approach which includes knowledge from the humanities and sciences. Moral norms mostly are quite general and abstract. For applying them to certain areas of action, situations or cases further reasoning is necessary. If several moral norms are in conflict, justification for their weighing must be given. Beside this, it must be methodologically possible and intended to integrate ethically relevant experiences and knowledge into ethical assessments; because many of them are so-called 'mixed' judgments, consisting of descriptions and assessments. Instead of only applying general norms to situations and institutional structures in a deductive way, an inductive way of integrating ethically relevant facts and human experiences is preferred.

Thus, in the following the categorical arguments for the moral status of the embryo will not be the only arguments for evaluating the field of stem cell research. Much more, the wider context of the practice of stem cell research will be examined and diverse supplement arguments will be presented.

17.2 Ethical Questions Raised by Research Using Human Embryos and Embryonic Stem Cells

In the past years, the framework as well as the goal of research using human embryos and embryonic stem cells have changed. Initially, embryo research took place within the context of reproductive medicine. For approximately three decades, use of human embryos for reproductive purposes had the 'side effect' of bringing about new insights into embryonic development and the interaction of genetic and epigenetic factors. The primary goal was to improve the cultures used for embryos created in-vitro and to increase the success of embryonic transfer and thus the success rates of assisted reproduction.

Since the late 1990s, in numerous countries, research on human embryos and embryonic stem cells has been conducted outside the context of reproductive medicine to an ever increasing degree in order to obtain new clinical-therapeutic applications and to promote pioneer research. After the sheep 'Dolly' was cloned, which is to say, after a genetically identical organism was created by transferring a somatic cell nucleus into a donor egg cell whose own nucleus had been removed, and the world experienced the first extraction of human embryonic stem cells (Thomson et al. 1998), hopes have been raised that it might become possible to create artificial human tissue. This could open up perspectives for reducing the immune rejection reaction of those receiving such tissue and for dealing with shortages of transplantable tissue or even entire organs. An additional possibility with clinical relevance is that pharmaceutical substances could be tested on human tissue and that in-vitro disease models could be constructed on the basis of such tests which would promote research on pathology and possible therapies. Furthermore, pioneer research efforts aim towards gaining additional insights into human developmental biology and genetic and epigenetic control mechanisms.

17.2.1 Ethical Problems Pertaining to the Extraction and/or Creation of Embryonic Stem Cells

There are currently three ways to extract or produce embryonic stem cells (Graumann and Poltermann 2004:22), but from an ethical standpoint, none of them are unproblematic.

First of all, it is possible to obtain embryonic stem cells from so-called surplus embryos which come into being as a result of in-vitro fertilization (IVF) in reproductive medicine. Secondly, embryonic stem cells can be cultivated from aborted fetuses. In recent years, the first method has been favored in particular. As opposed to creating human embryos for the sole purpose of conducting research, many view the use of 'surplus' embryos as unproblematic because they would otherwise be destroyed anyway.[1] The use of surplus embryos, which are obtained for the most part from reproductive-medical treatment of women, gives rise to the following objections, however:

As concerns both methods named so far, the self-purposefulness of the embryos is given due to the reproductive context: they have the status of potential human beings who could have developed an existence. By transferring them to a completely different context, i.e., that of research, the purposefulness becomes alienated because as a form of 'consuming' embryonic research, embryonic stem cell research uses embryos and stem cells for scientific purposes, destroying them in the process. Moreover it is quite questionable whether any 'free and informed consent' on the part of the biological parents or mothers suffices to put the embryos at the disposal of research, for this would entail a shift away from a general right to protection of human life to a concept of parentage which would be accompanied by an extensive right of disposal and ultimately a moral debasement of evolving human life (Habermas 2001:28 ff., 122 ff.). Habermas points out that in the past, such a right of disposal could only be applied to things, not to persons. Of course, there are good reasons for granting couples or women the right to make decisions concerning the well-being of a potential child and possible medical interventions to a certain degree in correspondence to their responsibilities towards it. The same holds for a woman's right to be respected as a moral subject in her decisions concerning reproduction, which is to say, her right to resist being compelled to continue a pregnancy, become sterilized or use birth contraception. And yet these rights cannot simply be transferred to the context of research without being challenged, because in this case no efforts to produce a viable human being or to promote his or her future well-being, to respect the bodily integrity of the woman involved or to improve her health through medico-therapeutic action are at issue.

Furthermore one must argue that the use of 'surplus' embryos could inaugurate a practice of willfully creating a surplus. A continuous demand for embryos in research, accompanied by the expectation of a continuous supply of embryos would influence the clinical practice of artificial fertilization. In light of this scenario, it is unlikely that models designed to avoid the creation of surplus embryos like those implemented in Austria would be embraced. It is certainly no coincidence that in Valencia, Spain, for example, a stem cell research center has been established adjacent to a fertility clinic, Spain being a country in which research on surplus embryos deriving from reproductive medicine is legally permissible (Bahnsen 2006). Apart from using 'surplus' embryos, this clinic requests that their patients donate egg cells. The institutionalized egg-cell donation program enables researchers to conduct cloning experiments, as these call for fresh egg cells (ones extracted from the ovaries directly) in large quantities. This example clearly shows that the problem of planned surpluses resulting from fertility treatment is inherent to any research which relies solely on 'surplus' embryos. If one takes this one step further and addresses the question of possible clinical applications, as does Ron Jones, the managing director of PPL Therapeutics, a private

---

[1] The Switzerland's Federal Law on Research with Embryonic Stem Cells from 2003 allows embryonic stem cells to be obtained from "surplus," approximately one-week-old IVF embryos.

bio-tech firm in the USA, a 'supply problem' soon arises (Schwägerl 2001). In order to provide a mass market of patients with replacement tissue, whose production depends on the use of embryos, the replenishment of supplies will probably not suffice, even if tens of thousands of embryos were to be available in Great Britain, Jones says. 'Many scientists are satisfied if their experiments are successful two or three times' (Schwägerl 2001), he notes, but as a businessman he sees himself forced to think in terms of production capacities on an industrial scale, for, as he argues, he naturally wants to sell his products to as many persons as possible.

The third way to obtain embryonic stem cells is through cloning, which entails importing the nucleus of a body cell into an egg cell whose own nucleus has been removed. No surplus embryos are used in this process, but rather new ones which have been created solely for the purpose of research. The cell nucleus of a somatic cell, which can be extracted from any human being, is introduced into an egg cell whose own nucleus has been removed, 99 percent of which then consists of genetic material from the body cell, approximately one percent of which consists of mitochondrial gene material from the 'donor' egg cell. Thus the human embryo which is created in this way is to a large extent the genetic clone of the organism which the body cell originated from. Through further cultivation processes embryonic stem cells can be obtained from this genetic clone.

The fact that through cloning, embryos are created for the sole purpose of research raises problems unlike those connected with the use of 'surplus' embryos. The cloning process itself – as well as so-called therapeutic cloning which is carried out later to produce immune-compatible replacement tissue – requires many egg cells. These must be extracted from women directly because it is very difficult to cryo-conserve egg cells. Apart from health risks (caused by hormone stimulation, superovulation and operative procedures) and the risks of self-exploitation and commercialization, problems pose themselves in cases where egg cell donation involves a potentially damaging procedure and potential donors are motivated by financial straits or egg cells are viewed as altruistic donations. It is also problematic to encourage women who subject themselves to IVF treatments due to unwanted childlessness to simultaneously participate in an egg-cell donation program. For one, hormone stimulation must be increased or repeated in order to generate additional egg cells, and this results in greater health risks. Secondly, psychological and ethical studies on human research sufficiently document just how difficult it is to guarantee voluntary and informed consent in the context of medical treatment, because many patients feel emotionally dependent on the goodwill of the physicians whose care they are in. Furthermore, medically assisted artificial fertilization usually involves additional fees or private financing, which might well encourage those involved to make 'barter deals'.

17.2.2 Impossibility of Separating the Reproduction and Research Dimension of the Cloning Procedure

The distinction between reproductive cloning on the one hand and non-reproductive, research-driven, or therapeutic cloning on the other has become well-established.

However, this conceptual differentiation is based on a form of pragmatic decisionism which is misleading in terms of ethical assessments. What happens to clones depends on decisions made by the researchers in question; if they want a clone to grow, it is implanted into the uterus of a woman, and if its stem cells are to be used for research, it is destroyed. Thus the terms of common usage refer to the applications for which the technology is to be used, which is to say, the intentions of the researchers involved, and not to the cloned embryo itself, its developmental potential or the issue of ethical need for protection. In part other ethical problems pose themselves as well if one artificially reproduces and fosters a living organism whose chromosome set is almost completely identical with that of the original rather than creating and 'consuming' cloned embryos solely for the purposes of research. In any case, terminology which suggests, from the very start, that non-reproductive cloning is less problematic than reproductive cloning must be criticized. Furthermore, the use of such terminology dissolves the connection between reproduction and embryo research on a conceptual level, but the fact is that stem cell research continues to rely on the context of reproductive medicine, for female egg cells are indispensible for procuring stem cell lines and this fact raises problems:

As was previously mentioned, the hormonal stimulation and superovulation which this process requires pose short- and long-term health risks for the women involved. And yet neither they nor they children profit from these serious medical risks. Furthermore, the issue of women's rights and gender must be taken into account (Haker 2005:140 ff.). Should women really be morally motivated to contribute in such a way to the common good, the future of national research practices or future patients? In research contexts, which are dominated by men, role-specific empathy and altruism are expected of women who only rarely fulfill decision-making functions in the political and the scientific arena and who very often put their generative and reproductive capacity at the disposal of their family and society at no cost (Schneider 2003:51 f.). Should an entire branch of research be established on this foundation?

A possible alternative might be to pay the donors, as is done in the case of assisted reproduction in many countries. In the past few years, some female researchers have argued in terms of women's rights, speaking out in favor of regulating egg cell donations on the model of property rights and contract law. The question that poses itself, however, is whether the problem can be solved by commercializing egg cell donations for use in other branches of research. By abandoning the principle that the human body may not be commodified, or the model of gratuitous donation, the tendency towards exploiting women from low-income groups or impoverished countries would increase.[2] Furthermore it would only be logical to expect a market for reproductive tissue to establish itself, a market on which egg cells and sperma are treated as replaceable raw materials completely divorced from their pro-creative biological and social significance.

---

[2] On the free market, egg cell "donations" which cost between 10,000 and 50,000 US-dollars are offered via Internet, for in the context of reproductive medicine, there is a scarcity of egg cell donations.

For one, it is questionable whether such practices would stop at non-reproductive cloning because endeavors are being made to obtain organ-compatible progeny. In the USA, some couples capable of reproducing receive IVF treatment in connection with pre-implantation diagnostics (PID) for the purpose of so-called HLA matching[3] in order to give birth to a child which could act as a bone marrow donor for a diseased sibling (Verlinsky et al. 2001; Kuliev et al. 2005). Furthermore, pioneer researchers are interested in the development of the genotype as a whole, the embryo produced in the laboratory and the phenotype, i.e. the living organism which later evolves, not only the early, embryonic stage of development. It goes without saying that these two 'slippery slope' arguments which warn against potential further developments fulfill the principle of plausibility more than they serve to justify proscriptions in any strict sense of the word. For the cause of the 'slippery' chain of events would not lie in the actions taken to perform research cloning itself, but rather in the expansion of conditions which would prove favorable for certain developments such as altruistic motives or scientific curiosity. At the same time, the analogies which are based on experiences call attention to future kinds of problems and plausible connections between actions.

17.2.3   The Researcher's Burden of Proof Concerning the Moral Status of Human Embryos and Embryonic Stem Cells

Embryos, and under certain circumstances embryonic stem cells possess the potentiality of becoming human beings, i.e., individuals. Even if their moral status is debatable, a consensus exists that in contrast to other cell systems, human embryos and embryonic stem cells constitute an ethical good. But what characterizes this ethical good and what need for protection derives from this status (Düwell 1998:34 ff.; Mieth 2002, 2006a)?[4]

The potential of a young embryo and perhaps also that of embryonic stem cells can be illustrated by the fact that we cannot imagine the process of becoming a human being without imagining ourselves as having been an embryo before nidation occurred, with this organism already constituting our bodiliness, our gender and our hereditary dispositions. An embryo has potential in the sense of the possibility and capacity to become a human being if it develops in accordance with its predisposition, i.e., if no action is taken or omitted which runs contrary to this predisposition. The argument that 'nature' does not implant all fertilized egg cells is not admissible

---

[3] HLA matching, i.e., human leukocyte antigen matching: through PID-selection of a suitable sibling bone marrow diseases of already existing children are to be treated through the transfer of healthy stem cells.

[4] For a differentiated line of argumentation concerning the moral status of the human embryo cf. Düwell (1998) and Mieth (2002), for the meaning of the "successive animation" in the tradition of the catholic church Mieth (2006).

here, for nature cannot be conceived of as a responsible subject. The argument which advocates worthiness of protection in regard to embryos is based on their continuity and identity with human beings qua persons and on their potential for being able to develop into one themselves. This provides no compelling reason for viewing embryos as carriers of moral rights directly in the way we do self-determined adults, for example. And yet the moral status of embryos establishes a worthiness of protection which at least prohibits any unspecific authorization of use for research. As concerns medical contexts in which moral rights of others, for example, women or parents, are directly affected, or research contexts with a potential for developing concrete therapies for patients who already have diseases, the situation would be somewhat different.

Within the framework of this investigation, the few remarks which I have made concerning the moral status of embryos and their worthiness of protection will have to suffice. I hope having made it clear that researchers who use human embryos must provide special justification for the procedures they adopt and the goals they pursue. The burden of proof lies on the side of the researchers; they must demonstrate concretely whether an individual human right meriting high regard exists which is opposed to the demand for the protection of human embryos against instrumentalization and the leveling of its moral status in the process of becoming a human being. And although worthiness of protection as it pertains to embryos differs from that enjoyed by self-determined adult human beings, it does not follow that embryos do not constitute human beings.

In ethical and legal evaluations, distinctions are sometimes made between human embryos and embryonic stem cells, with the former being designated as totipotent, while the latter are labeled as merely pluripotent (Swiss Federal Law 2005). If embryonic stem cells are not capable of developing into human beings but rather only have the capacity to develop into various types of cells – this being the assumption behind Switzerland's stem cell legislature and the position held by the German National Ethics Council (2001:42) – then, as the implicit conclusion goes, no substantial moral status can be attributed to these cells. And yet, the notion that embryonic stem cells are merely pluripotent does not reflect any unanimous scientific opinion, as is often claimed. In fact a consensus is insinuated which, in light of scientific uncertainty, does not hold.

For many years, a juridic dispute regarding the use of the terms 'totipotence' and 'pluripotence' as well as ongoing scientific uncertainty as to ways in which embryonic stem cells do indeed evidence totipotence have contradicted allegedly reliable biological facts. The term 'totipotence' designates, for one, the ability to form derivatives from all three germ layers, i.e., to develop into the various cell types which constitute the human body. Some authors use the term 'pluripotence' when referring to this ability. On the other hand, the term 'totipotence' is used to refer to the ability to develop all the different kinds of cells found in an embryo and thus the capacity to develop into a viable human being. In this case, if one speaks of the totipotence of a stem cell, what is meant is that it possesses all features of an egg cell necessary to initiate embryonic development. Among other things, this calls for information from the cell plasma. Theoretically, a stem cell which is totipotent in this sense of the word should be able to develop into a fully functioning blastocyst. Whether human embryonic stem cells do in fact fulfill this definition of the term 'totipotent' is a question which could only be answered by implanting some of them into a uterus. As concerns embryonic stem cells in mice and several other species of mammals, this has been proven to be the case (Nagy et al. 1990, 1993;, Beier 1999; Denker 2004, 2006).[5] Although no one has yet succeeded in creating viable embryos from embryonic stem cells of human beings alone, these can apparently develop into completely viable individuals if they are cultivated together with the nutrient cells (trophoblasts) which usually surround an embryo (National Ethics Council 2001:26).

Thus, it remains to be seen whether human embryonic stem cells can in and of themselves develop into viable human beings. In light of such uncertainties it would seem premature to view pluripotence as a given fact, however. Furthermore, ethical clarification concerning the question as to whether the aforementioned notion of totipotence encompasses the fulfillment of certain preconditions such as the existence of certain nutrient cells which are usually 'naturally' given is required. Any opposing assessment made on the basis of a scientifically unclarified matter without any further substantiation remains inconclusive. In fact there is substantial evidence to the contrary, namely that totipotence in an exhaustive sense of the word is given here, so that any attempts to deproblematize research with embryonic stem cells on the grounds that a lack of inherent potentiality prevents them from developing into human beings would seem to be unacceptable. Instead, an additional problem might pose itself, namely that by conducting research with embryonic stem cells a large number of genetically identical clones are created which have the potential for becoming human beings.

17.2.4 Explication, High Priority and Reachability of Therapeutic Research Goals

In response to the question as to what goals researchers pursue, the first one usually mentioned is a therapeutic goal, namely, to use research on young embryos to improve reproductive medicine, for example, to increase the success of IVF. One possible improvement concerns the nutrient medium, the aim being to succeed in cultivating embryos for longer periods of time so as to increase the accuracy of pre-implantation diagnostics. Such therapeutic goals require justification, however. At first sight, the possibility of being able to cultivate a human blastocyst for more than five days outside the uterus in order to gain more time for molecular-genetic examinations using more than one or two cells is quite appealing. And yet, if the

---

[5] Cf. for mice Nagy (1990, 1993), for an overview Denker (2004, 2006), and for a contra position, e.g. Beier (1999).

improvement of genetic pre-implantation diagnostics leads to the rejection of embryos while at the same time being looked upon as an unchallengeable goal of therapeutic research, what we find ourselves confronted with is a mixture of instrumental and ethical goals. Long-term goals such as the development of an artificial uterus also need to be legitimized in terms of the purposes for which it would be used as well.

The development of reproductive medicine, for example, illustrates the necessity for explicating and evaluating therapeutic research goals individually, for this field has been characterized by a dual set of objectives ever since it began to establish itself in the early 1970s. Looking back at the development of IVF, Edward Brown, the 'father' of the first test-tube baby, attests that the desire to gain new insights into embryonic development and molecular genetics guided researchers' actions to the same degree as did the prospect of developing means of treating infertility (Brown 1998).

In the meantime, embryonic and stem cell research is used to pursue therapeutic goals in other medical disciplines as well, such as the production of replacement tissue and, under certain conditions, even transplantable organs. In particular, perspectives for developing therapies for diseases which are as yet incurable or hard to treat such as diabetes, Morbus Parkinson and Alzheimer establish a necessity for procuring and utilizing embryonic stem cells. Within the scientific community, notions differ concerning the reachability and quality of final therapeutic goals as well as the ability or inability to substitute the means used to reach these goals. For one, the prospect of clinically applicable cell replacement therapies is controversial; one might consider the clinical results of animal models for Morbus Parkinson, for example. Other forms of therapy even seem to offer more promising alternatives for treating this disease, such as electro-stimulation (Bentele 2007).

In attempts to justify research with embryonic stem cells, the argument that it could help to improve therapies and techniques is sometimes used strategically, with patients being instrumentalized by unrealistic promises of curability. The microbiologist and Parkinson patient Hans Zähner reproaches researchers as follows: 'We patients feel betrayed and misused. Betrayed, because hopes were raised by those who could know, who must know that they cannot be fulfilled. Misused because we are deployed to combat resistance against stem cell research' (Zähner 2002:72).

And the politically inspired coinage of new terms such as 'therapeutic cloning' helps blur the goals which have been set. In the case of cloning procedures currently in use, it would be more apt to speak of 'research cloning,' as the development of any conceivable clinical application remains a thing of the far future. In particular this concerns the prospect of being able to produce transplantable organs. As yet we have no knowledge of the temporal and spatial degree of coordination which comes into play in the differentiation and growth processes of various cells and tissue within any given organ, nor can we simulate it in any laboratory.

Ethical assessments come to different conclusions in cases where patients who already suffer from diseases have concrete hopes for help provided by new therapies as opposed to cases where researchers can only formulate vague notions concerning possible clinical applications. What is termed therapeutic research must be scrutinized in terms of what realistic goals are pursued to the benefit of which concrete individuals in possession of certain rights and which kind of medical help such individuals are entitled to (Gewirth 1978). Until it is possible to clarify whether such medical performance is really to be viewed as therapeutic – in the sense of providing cures for concrete patients – the argument that research on human embryos will result in an improvement of therapeutic options may not be brought into the equation to an unlimited degree. For in the case of an assessment which weighs morally relevant goods and moral rights against each other one must distinguish between therapeutic research with concrete applicability, i.e., research which offers realistic success for certain individuals, and research whose clinical applicability is questionable or merely anticipated as an option for the far future.

17.2.5   Prospects for Success and Lack of Alternatives in Research Using Embryos and Embryonic Stem Cells

As concerns the creation of replacement tissue and organs on the basis of cloning, great uncertainty prevails as to whether and when clinically applicable therapies can be developed. Some researchers such as the brain and stem cell researcher Wiestler even deem it to be improbable that this will ever happen. As he contends, the genetic programs of all cells obtained in this way evidence defective control mechanisms, arguing that it would be completely unacceptable to implant a cell whose genetic blueprint had been destroyed (Wiestler 2004). This problem fundamentally calls this technology into question, for not only are hundreds of egg cells required in order to create a viable mammal clone, and to obtain, in turn, a few intact embryos and ultimately one or two stem cell lines. There are also indications that animal experiments motivated by this goal carry a high risk potential in terms of deformities, tumors and accelerated aging.

The hope is that research cloning will lead to the production of immune-compatible organ cells. And yet many questions remain unanswered. How can it be proven that the transfer of cells causes the diseased organ to begin functioning again rather than resulting in damage of the implanted cells as well? Furthermore, the age of the stem cell deriving from the original organism is apparently transferred via the stem cell lines, as has been shown in animal experiments using the 'Dolly method.' How can we deal with the problem of cell aging? Has the problem concerning immune defense actually been solved in light of the fact that genetic material stemming from the foreign egg cell whose nucleus has been removed also influences the cloned cell material? Thus there is complete uncertainty as to how the embryonic stem cells which have differentiated into individual body cells would behave in the human body. In light of this, Wiestler sees the relevance of clone experiments, which according to his assessment can for the most part be conducted on animal cells, as limited to basic research: 'Perhaps they help us to understand which factors in the cell sap of an egg cell serve to bring a cell nucleus back to a very early stage of development (Wiestler 2004).'

Not only can we ascertain that cloning technology is not necessarily suitable as a point of departure for the clinical therapies researchers aim to develop, but that other forms of basic research have not yet been pursued which might prove to be relevant for the treatment of those diseases named in connection with research cloning. Cell replacement therapy would not enable us to cure Diabetes mellitus or Morbus Alzheimer, for example (Wodarg 2004). For in the widest sense of the word, Diabetes mellitus is an auto-immune disease, and thus replacement cells would presumably be affected in the same way that the body's own cells are. Until we have gained an understanding of the pathological process it is unlikely that much can be done by implanting replacement tissue repeatedly. And problems concerning tissue rejection and tumor formation have not been solved yet either (President's Council on Bioethics 2004:135 ff.). Alzheimer's disease, which entails damage of the brain caused by protein deposits, is a similar case.

In order to justify conducting research with embryonic stem cells it does not suffice to raise general hopes of new forms of therapy. Instead, already existing therapies and prospective therapies as well as their future developmental potential must be analyzed. Furthermore, it would be necessary to show whether and which clinical symptoms of a disease are so serious that they warrant resorting to ethically sensitive goods for research purposes under certain conditions.

At present, cell or even organ therapy cannot be looked upon as the remedy for the future and for this reason, prospects of its future feasibility do not serve as justification for embryonic research. Therapy goals which claim high priority in research must be scrutinized in terms of their realizability before they can carry any weight as legitimation for the utilization of human embryos. The sciences must provide concrete evidence and also show that no other means and methods exist which are less problematic from an ethical perspective. Moreover, the problem-solving rule must be taken into consideration, which says that new developments should not create problems more serious than the ones they claim to solve. The 'problem balance sheet' for somatic nucleus transfer, for example, does not necessary speak in favor of its clinical application.

As soon as researchers stop setting their sights on clinical-therapeutic applications and focus, instead, on basic scientific insights, ethical goods and goals must be weighed differently. From an ethical perspective, the issue as to whether human embryos and stem cells may be used to achieve far-reaching goals of pioneer research or not is extremely controversial. For in this case it is not possible to argue that there is a lack of alternatives. This argument does not hold for any general gain in knowledge, but rather for research on therapies and the development of techniques and technologies which promise to cure or alleviate diseases – and, one must add, in a justifiable way. Equally inadmissible is the argument that, in the interests of new insights, research should not be constrained in any way. Thus pioneer research which explicitly defines itself as such must be evaluated separately. The expectations placed in embryonic stem cell research are that, among other things, new insights into fundamental mechanisms of cell programming, the transition from the geno- to the phenotype and embryonic development could be gained through it. This is not objectionable, and yet in pioneer research, means and methods for which a moral consensus exists should be favored – particularly in light of the fact that their outcome is uncertain. Thus it would be possible, for example, to continue working with animal models, animal embryos and embryonic stem cells from animals as well as human adult stem cells with an aim towards gaining a better understanding of fundamental mechanisms which regulate the programming and reprogramming of somatic stem cells or the immune system. Were research on human embryos and embryonic stem cells to be prohibited, this would not bring pioneer research in the area of molecular genetics and embryology to a halt. Were research funds to be invested in creative, farther-reaching efforts to find ethically unproblematic alternatives, this could help to establish new areas of research.

17.2.6 Ethical Assessment and Procedural Limitations

For several years, procedural limitations at the national and international level have been set to fulfill the demands for protection of embryos. These include notification requirements, justification requirements and licensing through a commission, for example. Legally speaking, if this practice continues to be pursued, this will mean a shift in German, and, to a large extent, European interpretation of law in the direction of the Anglo-Saxon legal system.

Ethically speaking, if decisions concerning important ethical issues are delegated to commissions and procedures, this means that the protection of embryos will no longer hold for the integrity of each individual embryo, this implicating, in turn, a rejection of what has in the past been deemed a high priority (Mieth 2006b). For regulatory procedures cannot fundamentally change questionable research practices. What lies behind such procedures is instead the ethically questionable assumption that a good end justifies poor means if these are used 'sparingly and carefully.'

Strong pragmatic interests or formal procedures, which tend to produce compromises or follow majorities do not suffice, if fundamental normative questions have to be ethically assessed. The burden of proof is on the side of the scientific research, if sensitive human goods and values are touched or even at stake.

References

Auer, Alfons (1971). *Autonome Moral und Christlicher Glaube*, Düsseldorf, Germany: Patmos.

Bahnsen, Ulrich (2006). "Neubeginn im Klonlabor. Die Versuche gehen weiter", *Die Zeit*, May 24.

Beier, Henning M. (1999). "Definition und Grenze der Totipotenz: Aspekte für die Präimplantationsdiagnostik", *Ethik in der Medizin*, 11, S 23–37.

Bentele, Katrin (2007). *Ethische Aspekte der regenerativen Medizin am Beispiel von Morbus Parkinson*, Münster, Germany: LIT-Verlag.

Brown, Robert G. Edwards (1998). "Introduction of IVF and its ethical regulation". In: Hildt, Elisabeth and Mieth, Dietmar (eds.). *In Vitro Fertilisation in the 1990s*, Aldershot, England: Ashgate, pp. 3–18.

Denker, Hans-Werner (2004). "Early human development: new data raise important embryological and ethical questions relevant for stem cell research", *Naturwissenschaften*, 91(1), 1–21.

Denker, Hans-Werner (2006). "Potentiality of embryonic stem cells: an ethical problem even with alternative stem cell sources", *Journal of Medical Ethics*, 32, 665–671.

Düwell, Marcus (1998). "Ethik der genetischen Frühdiagnostik – eine Problemskizze". In: Düwell, Marcus and Mieth, Dietmar (eds.). *Ethik in der Humangenetik: die neueren Entwicklungen der genetischen Frühdiagnostik aus ethischer Perspektive*, Tübingen, Germany: Francke, pp. 26–50.

Gewirth, Alan (1978). *Reason and Morality*, Chicago, IL: University of Chicago Press.

Graumann, Sigrid and Poltermann, Andreas (2004). "Klonen: ein Schlüssel zur Heilung oder eine Verletzung der Menschenwürde?", *Aus Politik und Zeitgeschichte*, 23–30.

Habermas, Jürgen (2001). *Die Zukunft der menschlichen Natur: auf dem Weg zu einer liberalen Eugenik?*, Frankfurt/M., Germany: Suhrkamp.

Haker, Hille (2005). "Ethische Aspekte der embryonalen Stammzellforschung". In: Bender, Wolfgang, Hauskeller, Christine and Manzei, Alexandra (eds.). *Grenzüberschreitungen: Kulturelle, religiöse und politische Differenzen im Kontext der Stammzellenforschung weltweit*, Münster, Germany: Agenda Verlag, pp. 127–154.

Kuliev, Anver, Rechitsky, Svetlana, Tur-Kaspa, Ilan, and Verlinsky, Yury (2005). "Preimplantation genetics: improving access to stem cell therapy", *Annals of the New York Academy of Sciences*, 1054, 223–227.

Mieth, Dietmar (2002). "Forschung an embryonalen Stammzellen". In: Mieth, Dietmar (ed.). *Was wollen wir können?*, Freiburg im Breisgau, Germany: Herder, pp. 241–259.

Mieth, Dietmar (2006a). "Stem cells: The ethical problems of using embryos for research", *The Journal of Contemporary Health Law and Policy*, 22, 439–447.

Mieth, Dietmar (2006b). Embryonale Stammzellen – die spezielle Fortschrittsverantwortung, unpublished contribution, Germany: University of Tübingen.

Nagy, Andras, Gocza, Elena, Diaz, Elizabeth Merentes et al. (1990). "Embryonic stem cells alone are able to support fetal development in the mouse", *Development*, 110, 815–821.

Nagy, Andras, Rossant, Janet, Nagy, Reka, Abramow-Newerly, Wanda, and Roder, John C. (1993). "Derivation of completely cell culture-derived mice from early-passage embryonic stem cells", *Proceedings of the National Academy of Sciences of the United States of America*, 90, 8424–8428.

National Ethics Council (Nationaler Ethikrat) (2001). Zum Import menschlicher embryonaler Stammzellen, Saladruck, Berlin: Stellungnahme.

Schneider, Ingrid (2003). "Gesellschaftliche Umgangsweisen mit Keimzelle: Regulation zwischen Gabe, Verkauf und Veräußerlichkeit". In: Graumann, Sigrid and Schneider, Ingrid (eds.). *Verkörperte Technik, entkörperte Frau: Biopolitik und Geschlecht*, Frankfurt, Germany: Campus, pp. 41–65.

Schwägerl, Christian (2001). "Strohhalme für die Ethik - Embryonen, die keine sind: Die Forschung hat neue Ideen", *Frankfurter Allgemeine Zeitung*, Oct. 20.

Swiss Federal Law on Research with Embryonic Stem Cells (Stem Cell Research Law, STFG) from Dec. 19, 2003, put into effect in Jan. 2005.

The President's Council on Bioethics (2004). Chapter "Stem cell therapy for type-1 diabetes?", *Monitoring Stem Cell Research*, Washington, DC.

Thomson, James A., Itskovitz-Eldor, Joseph, Jones, Jeffrey M. et al. (1998). "Embryonic stem cell lines derived from human blastocysts", *Science*, 282, 1145–1147.

Verlinsky, Yury, Rechitsky, Svetlana, Schoolcraft, William, Strom, Charles et al. (2001). "Preimplantation diagnosis for Fanconi anemia combined with HLA matching", *JAMA*, 285, 3130–3133.

Wiestler, Otmar (2004). "Teure Irrwege. Interview mit Otmar Wiestler zu Bedeutung der Klon-Embryos für die Forschung", *Süddeutsche Zeitung*, Feb. 14/15, 2004.

Wodarg, Wolfgang (2004). "Die koreanische Lüge. Was die Klon-Forscher verschweigen", *Süddeutsche Zeitung*, April 14/15, 2004.

Zähner, Hans (2002). "Interview mit Katrin Bentele: Hoffen und Bangen. Die Versprechungen der Stammzellforschung aus Sicht eines Parkinson-Patienten". In: Dietrich, Julia (ed.). *Humane Genetik?*, Seelze, Germany: Friedrich, pp. 71–72.

Index